Lecture Notes in Economics and Mathematical Systems

391

Claudia Keser

Experimental Duopoly Markets
with Demand Inertia

Game-Playing Experiments
and the Strategy Method

Springer-Verlag

Berlin Heidelberg New York
London Paris Tokyo
Hong Kong Barcelona
Budapest

Author

Dr. Claudia Keser
Institut für Gesellschafts- und Wirtschaftswissenschaften
Universität Bonn
Adenauerallee 24-42
W-5300 Bonn 1, FRG

ISBN-13: 978-3-540-56090-6   e-ISBN-13: 978-3-642-48144-4
DOI: 10.1007/ 978-3-642-48144-4

© Springer-Verlag Berlin Heidelberg 1992

Typesetting: Camera ready by author/editor
42/3140-543210 - Printed on acid-free paper

To my mother

(...) Livet maa forstaaes baglænds. Men (...) det maa leves forlænds.*

S¢ren Kierkegaard

*   Life can only be understood backwards but it has to be lived forwards.

CONTENTS

1. INTRODUCTION

This report portrays the results of experimental research on dynamic duopoly markets with demand inertia. Two methods of experimentation are studied: game–playing experiments where subjects interact spontaneously via computer terminals, and computer tournaments between strategies designed by subjects. The principal aim of this study is the understanding of boundedly rational decision making in the dynamic duopoly situation.

1.1 Motivation

The experiments examine a multistage duopoly game where prices in each period are the only decision variables. Sales depend on current prices and also on past sales (demand inertia). Applying the game–theoretic concept of subgame perfect equilibrium, the game is solved by backward induction. The result is a uniquely determined system of decision rules.

However, we can hardly expect that human beings behave according to the equilibrium strategy of this game. It is unlikely that subjects are able to compute the equilibrium. And even if a subject is able to compute it, he might not make use of this knowledge. Only if he expects the others to behave according to the equilibrium, it is optimal for him to play the equilibrium strategy. We have evidence from several earlier experimental studies on oligopoly markets that, even in less complex oligopoly situations where the equilibrium solutions are very easy to compute, human behavior often is different from what is prescribed by normative theory.[1]

Normative theory is based on the concept of ideal rationality. However, human capabilities impose cognitive limits on rationality. Furthermore, there are motivational limits on rationality. A person may know very well what the equilibrium solution is but feel reluctant to behave accordingly. In the dynamic duopoly situation, a person might expect the other person to behave differently

---

[1] See for example Friedman and Hoggatt (1980), Alger (1980), Selten, Mitzkewitz and Uhlich (1988).

from the equilibrium solution. His decision might be guided by the hope of gaining, by cooperation with the other, higher profits than in the equilibrium.

For this reason, we suspect that actual human behavior in the duopoly situation with demand inertia is different from what is prescibed by the equilibrium solution. A very popular reaction would be to construct a new theory based on Baysian methods. However, Baysian theory is also based on very strong rationality assumptions. Baysian optimization exceeds human capabilities as well.

In order to develop a descriptive theory of boundedly rational behavior, we need to get insight into the structure of human decision making by conducting experiments. The main purpose of this study is to analyze the behavior of boundedly rational subjects in a dynamic duopoly situation with demand inertia.

The experimental analysis of the duopoly situation with demand inertia is an interesting enlargement of on–going research on experimental oligopoly markets. The dynamic structure of the game allows to study aspects of human decision making which do not have an effect in less complex oligopoly situations.

1.2   The experimental methods

Two methods of experimentation have been applied. I started with a series of game–playing experiments. This is the more traditional method of experimentation: Subjects are instructed about a game. Then, they play this game against each other by direct interaction via computer terminals. They are paid according to their success in the game so that they have an economic incentive to do well.[2]

Results from several experimental studies of oligopoly markets suggest that the subjects' experience with the game matters.[3] Thus, in order to explore the influence of experience, I let each subject play the duopoly game twice.

---

[2] The game–playing experiments were conducted in the Laboratory of Experimental Economics at the University of Bonn. Subjects were sixty students from the Department of Law and Economics.

[3] See for example Stoecker (1980), Benson and Faminow (1988).

The second experimental method which I applied is based on the strategy method (Selten (1967)): Subjects get the opportunity to become experienced in playing a specific game, and then design strategies for this game in flow–chart–form. A strategy is a complete and exact behavioral plan, assigning a decision to every situation which may arise. I applied this strategy method in computer tournaments: Subjects' strategies have been translated into computer programs so that they can play against each other in computer simulations. Thus, not only the structure of the strategies may be analyzed but also their success when they interact with each other. This is of great importance for the subjects' motivation.

The computer tournament with strategies has been organized as an international strategy tournament with a general invitation to participate. Participants were recruited by the sending out of a "call for participation" in an information brochure describing the tournament rules.

This way of applying the strategy method leads to the problem that it is difficult to provide the subjects with the necessary experience in playing the game for which they are to design the strategy. This is of particular importance for a more complex game like the dynamic duopoly. I tried to overcome this difficulty by including into the information brochure some results of the game–playing experiments. Furthermore, a second tournament round was announced already in the information brochure. To the second tournament, only the participants with the experience of the first tournament were permitted.

The great advantage of the strategy method in comparison with the game–playing method of experimentation is that it makes the strategies of the subjects directly visible. In game–playing experiments we can observe actual plays but the subjects' strategies remain hidden. Strategies convey more information about the structure of subjects' decision making in the considered game situation. The strategy method, however, also causes comment. A very important criticism is that the strategy method might not fully reveal the structure of actual behavior. Subjects are forced to design a complete strategy. However, we cannot suppose that subjects have a complete behavioral plan in mind when they play a game spontaneously. We suppose, nevertheless, that the strategy method reveals to a large extent the motivational forces which influence the spontaneous behavior.

In this study, I compare the results of both experimental methods in order to see what distinguishes spontaneous behavior from a strategy. So far, no such comparison is known in the literature.

1.3 Structure of the book

The model and some theoretical features are presented in Chapter 2. Chapter 3 describes the organization of the game–playing experiments and the strategy tournaments. In Chapter 4, we see the main results of the game–playing experiments: Section 4.1 describes the average behavior while Section 4.2 analyzes the individual markets. Chapter 5 portrays the results of the strategy tournaments: In Section 5.1 some general results of the tournaments are described. Section 2 is the central section of Chapter 5. It analyzes the structure of the strategies. Section 5.3 examines what determines the success of a strategy. Section 5.4 reports the results of an evolutionary process based on the strategies of the second tournament. Chapter 6 compares the results of game–playing experiments and strategy tournaments. In Chapter 7, the results of this study are compared with those of related studies. A summary of the main results is given in Chapter 8.

2.   THE MODEL

The model is one of a multistage price–setting duopoly with demand inertia. The notion "demand inertia" refers to a dynamic relationship between present sales and past sales. This relationship arises because sales depend both on current prices and past sales. Imagine customers tending to stay with the same seller but also tending to change from high price sellers to low price sellers. The situation is modelled as a non–cooperative game.

The origines of the non–cooperative game are in Selten (1965). In this seminal paper Selten introduced the notion of subgame perfect equilibrium to solve a model of an oligopoly with demand inertia. My study considers a duopoly with fixed parameters belonging to the class of games analyzed by Selten. The parameters are chosen such that the game has a unique subgame perfect equilibrium solution. In this chapter we will see the specific model as it was presented to the subjects of the experiments, its subgame perfect equilibrium solution, conditions for Pareto optimality, and a specific cooperative solution.

2.1   The game

There are two firms in a market, offering a homogeneous product over 25 periods. The only decision variable of each firm is its price in each period.

Let $x_i{}^t$ be the quantity sold by firm i in period t. This quantity is assumed to depend linearily on the firm's own price in period t, $p_i{}^t$, and on a variable $D_i{}^t$ which is called the firm's "demand potential" in period t:

$$x_i{}^t = D_i{}^t - p_i{}^t \qquad \text{i=1,2 ; t=1,..,25} \tag{1}$$

In the first period, the demand potential is 200 for each firm:

$$D_i{}^1 = 200 \qquad \text{i=1,2} \tag{2}$$

In the following periods, a firm's demand potential is determined by its former demand potential and the price difference of the two firms in the previous period:

$$D_i^t = D_i^{t-1} + \frac{1}{2}(p_j^{t-1} - p_i^{t-1}) \qquad i=1,2 \ ; \ j=3-i \ ; \ t=2,..,25 \qquad (3)$$

We see that if a firm in period $t-1$ has offered its product at a lower price than the opponent firm, then its demand potential for period t increases. The demand potential of the other firm decreases by the same amount. Obviously, the average demand potential in the market remains constant over all periods, but what may change from one period to the next is how the two firms share the total demand potential of the market.

The firms are assumed to produce exactly what they can sell. Production costs result from constant unit cost c_i which are

$$c_1 = 57 \ ; \qquad c_2 = 71 \qquad\qquad\qquad (4)$$

We call firm 1 the "low cost" firm and firm 2 the "high cost" firm.

The profit of firm i in period t is given by

$$g_i^t = (p_i^t - c_i)x_i^t \qquad\qquad i=1,2 \ ; \ t=1,..,25 \qquad (5)$$

This profit is credited after each period on an account which yields an interest rate of 1% per period. Thus, firm i's total profit after 25 periods is given by

$$G_i^{25} = \sum_{t=1}^{25} (1.01)^{25-t} g_i^t \qquad\qquad i=1,2 \qquad (6)$$

In the following, we shall call this firm i's "long-run profit".

The procedure of the game is the following: In each period t the two firms in the market independently decide about the prices of the actual period t. Both firms are completely informed about the rules of the game. They not only know the formulas of the model but also all the parameters. Furthermore, they have complete information of everything that has happened in their market so far. Thus, they know their actual demand potentials when they fix their prices. They also know that the game ends after 25 periods. Each firm wants to maximize its long–run profit.

The firms have to obey two <u>constraints</u> when they fix their prices. First, they are not allowed to choose prices above their actual individual demand potentials. This would yield negative sales. Secondly, to prevent dumping, they are not allowed to set prices below their own unit cost of production. However, in the event that a firm's demand potential should have fallen below its unit cost, the price is automatically fixed at its demand potential so that the firm does not sell anything. Thus, neither a firm's demand potential nor its short–run profits can ever become negative.

2.2 The subgame perfect equilibrium solution

This section describes the subgame perfect equilibrium solution of the game presented above. The solution follows directly form a translation of the results by Selten (1965) into the specified game situation of my study.

Selten first analyzes an extended game without any positivity constraints on quantities and prices. In the case of finitely many periods, the extended game has a unique subgame perfect equilibrium. Selten derives sufficient conditions on the parameters under which the subgame perfect equilibrium path of the extended game does not violate the positivity restrictions. These conditions are satisfied for the experimental game. Strictly speaking, the subgame perfect equilibrium solution described in the following is the uniquely determined subgame perfect equilibrium point of the extended game. It seems to be a plausible conjecture that an exact analysis of the game actually played would not yield anything essentially different. It is unlikely that it ever pays for one firm to push the other one into a situation in which it cannot follow the prescriptions of the subgame perfect equilibrium solution since it violates the positivity restrictions. The experience with my strategy tournaments shows that aggressive low cost strategies which try to destroy the high cost firm's ability to take prices above its unit cost are far from being best replies to the equilibrium solution.

As we shall see, the price path of the subgame perfect equilibrium solution never comes close to a violation of the positivity restrictions. This shows that the equilibrium solution is stable against small deviations. Even if a full analysis of the game with its positivity restrictions remains an open question, it seems to be justified to look at the subgame perfect equilibrium solution as a normative base line with which experimental results can be meaningfully compared.

The price paths in the subgame perfect equilibrium solution are given by

$$p_i^{t*} = a^t(D_i^t - 200) + b^t + k^t \frac{(c_i - c_j)}{2} \qquad i=1,2; \; j=3-i; \; t=1,..,25 \qquad (7)$$

where the parameters a^t, b^t and k^t are independent of the actual individual demand potential and the individual production cost. They can be calculated backwards from the last period to the first by a recursive system. As they depend only on the interest rate, the average production cost and the market demand potential, they are the same for both firms in the market.

The parameters can be calculated by the following recursive system, where we use A^t, B^t, K^t, Y^t, and z^t as auxiliary variables:

Given the values for the last period

$$
\begin{array}{ll}
a^{25} = 0.5 & A^{25} = 0.25 \\
b^{25} = 132 & B^{25} = 68 \\
k^{25} = 0.5 & K^{25} = -0.5 \\
Y^{25} = \dfrac{1}{4.04} &
\end{array}
\qquad (8)
$$

we can compute for t = 24 down to 1

$$a^t = \frac{1 - Y^{t+1}}{2 - Y^{t+1}}$$

$$z^t = \frac{1}{2.02}(1.5 - a^t)$$

$$A^t = 0.25 + z^t(1 - a^t)A^{t-1}$$

$$b^t = 132 - \frac{1}{4.04}B^{t+1} \tag{9}$$

$$B^t = 68 + z^t B^{t+1}$$

$$k^t = \frac{1 - \frac{1}{2.02}K^{t+1}}{2 - Y^{t+1}}$$

$$K^t = -0.5 + z^t(K^{t+1} - 2A^{t+1}k^t)$$

$$Y^t = \frac{1}{4.04} + z^t(1 - a^t)Y^{t+1}$$

The equilibrium price paths and the resulting demand potentials for the two firms are given in Figure 2.1. We see that in the first periods the high cost firm loses part of its demand potential by setting higher prices than the opponent firm. Then, the prices of both firms assimilate: The high cost firm decreases while the low cost firm increases its price. In the middle part of the game, the firms' prices increase together but by so little that it is hardly visible in the figure. During these periods, the difference between the market shares of both firms remains constant: The high cost firm remains with a demand potential significantly lower than its opponent's. Towards the end, when long–run profit maximization becomes less and less important, the prices of both firms increase drastically. The prices of the last period are the short–run monopoly prices.[4] The resulting profits are 136 728 for the low cost firm and 61 498 for the high cost firm.

---

[4] See equation (21) below.

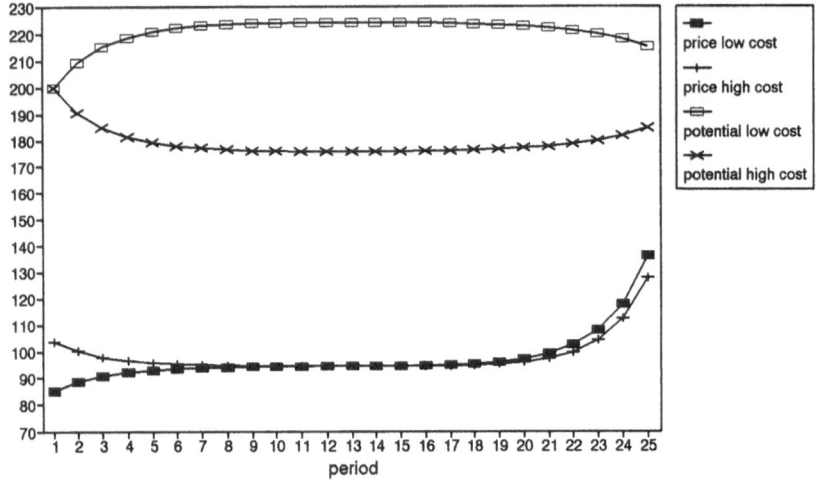

Figure 2.1: The subgame perfect equilibrium solution

2.3 Pareto optimality

We are interested in the potential scope for cooperation in the dynamic duopoly and thus the characterization of Pareto optimal outcomes. Pareto optimal outcomes have the property that the profit of one player cannot be increased without reducing the profit of the other player.

To characterize Pareto optimal outcomes of the dynamic game, we maximize backwards, for all $t = 25,...,1$, the weighted sum of discounted profits of the subgame starting in period t:

$$M^t = \lambda \sum_{k=t}^{25} q^{k-t} g_1^k + (1-\lambda) \sum_{k=t}^{25} q^{k-t} g_2^k \tag{10}$$

where $q = \dfrac{1}{1.01}$ is the discount factor

and $\quad 0 < \lambda < 1$

Maximizing the weighted sum of profits for the last period only, we get

$$M^{25,po} = \alpha_0{}^{25} + \alpha_1{}^{25} V^{25} + \alpha_2{}^{25} (V^{25})^2 \tag{11}$$

where $V^t = \dfrac{D_1{}^t - D_2{}^t}{2}$ $\tag{12}$

is called the "advantage of firm 1" in period t and

$$\alpha_0{}^{25} = \lambda\left(100 - \tfrac{c_1}{2}\right)^2 + (1-\lambda)\left(100 - \tfrac{c_2}{2}\right)^2$$

$$\alpha_1{}^{25} = 200\lambda - 100 - \frac{(\lambda c_1 - (1-\lambda)c_2)}{2} \tag{13}$$

$$\alpha_2{}^{25} = 0.25$$

Given this solution for the last period, we can write M^t for period $t = 24,23,...,1$ as

$$M^t = \lambda g_1{}^t + (1-\lambda)g_2{}^t + q[\alpha_0{}^{t+1} + \alpha_1{}^{t+1} V^{t+1} + \alpha_2{}^{t+1}(V^{t+1})^2] \tag{14}$$

Maximization of this expression yields similarily to (11)

$$M^{t,po} = \alpha_0{}^t + \alpha_1{}^t V^t + \alpha_2{}^t (V^t)^2 \qquad\qquad t=1,...,24 \qquad\qquad (15)$$

where $\alpha_0{}^t$, $\alpha_1{}^t$ and $\alpha_2{}^t$, which are independent of V^t, can be computed for given $\alpha_0{}^{t+1}$, $\alpha_1{}^{t+1}$, $\alpha_2{}^{t+1}$ from

$$\alpha_0{}^t = \left[\frac{200-\gamma}{2}\right]^2 + \frac{(\beta_3{}^t)^2}{\beta_1{}^t} + q\alpha_0{}^{t+1}$$

$$\alpha_1{}^t = (200+\gamma)(\lambda-\tfrac{1}{2}) + 2\frac{\beta_2{}^t\beta_3{}^t}{\beta_1{}^t} + (1-\lambda)c_2 - \lambda c_1 + q\alpha_1{}^{t+1} \qquad (16)$$

$$\alpha_2{}^t = (\lambda-\tfrac{1}{2})^2 + \frac{(\beta_2{}^t)^2}{\beta_1{}^t} + q\alpha_2{}^{t+1}$$

where γ, $\beta_1{}^t$, $\beta_2{}^t$ and $\beta_3{}^t$ are auxiliary variables to simplify notation:

$$\beta_1{}^t = -(q\alpha_2{}^{t+1} - 4\lambda(1-\lambda))$$

$$\beta_2{}^t = q\alpha_2{}^{t+1} - 2\lambda(1-\lambda) \qquad\qquad (17)$$

$$\beta_3{}^t = \tfrac{1}{2}q\alpha_1{}^{t+1} - \lambda(1-\lambda)(c_1-c_2)$$

$$\gamma = \lambda c_1 + (1-\lambda)c_2$$

With these parameters, and knowing that

$$V^1 = 0$$

$$\qquad\qquad (18)$$

$$V^t = V^{t-1} + \tfrac{1}{2}(P_2{}^{t-1} - P_1{}^{t-1}) \qquad\qquad t=2,...,25$$

we can compute, for given λ, the Pareto optimal prices for $t = 1,...,24$ as

$$P_1{}^t = \tfrac{1}{2}(200 + \gamma + (2\lambda{-}1)V^t) - 2(1{-}\lambda)\,\frac{\beta_2{}^tV^t + \beta_3{}^t}{\beta_1{}^t}$$

$$\tag{19}$$

$$P_2{}^t = \tfrac{1}{2}(200 + \gamma + (2\lambda{-}1)V^t) + 2\lambda\,\frac{\beta_2{}^tV^t + \beta_3{}^t}{\beta_1{}^t}$$

and for the last period as

$$P_1{}^{25} = \tfrac{1}{2}(200 + V^{25} + c_1)$$

$$\tag{20}$$

$$P_2{}^{25} = \tfrac{1}{2}(200 - V^{25} + c_2)$$

For $c_1 = 57$ and $c_2 = 71$, the resulting long–run profits $G_i{}^{25}(\lambda)$ turn out to be positive for both players in a range of λ between about 0.438 and 0.462. Table 2.1 reports the Pareto optimal profits for some values of λ in this range. The frontier of maximal possible profits is shown in Figure 2.2. This figure also indicates the long–run profits resulting for both firms in the subgame perfect equilibrium solution. They are quite far below the Pareto frontier.

Table 2.1

List of Pareto optimal profits for some values of λ (.438 < λ < .462).

λ	Pareto optimal profit of	
	low cost firm	high cost firm
.439	4 091	228 582
.440	22 833	213 887
.441	40 647	199 862
.442	57 612	186 451
.443	73 799	173 604
.444	89 272	161 273
.445	104 088	149 417
.446	118 303	137 997
.447	131 963	126 978
.448	145 113	116 327
.449	157 794	106 014
.450	170 042	96 013
.451	181 893	86 298
.452	193 378	76 844
.453	204 526	67 631
.454	215 364	58 637
.455	225 917	49 844
.456	236 209	41 235
.457	246 260	32 792
.458	256 092	24 501
.459	265 723	16 346
.460	275 171	8 314
.461	284 453	392

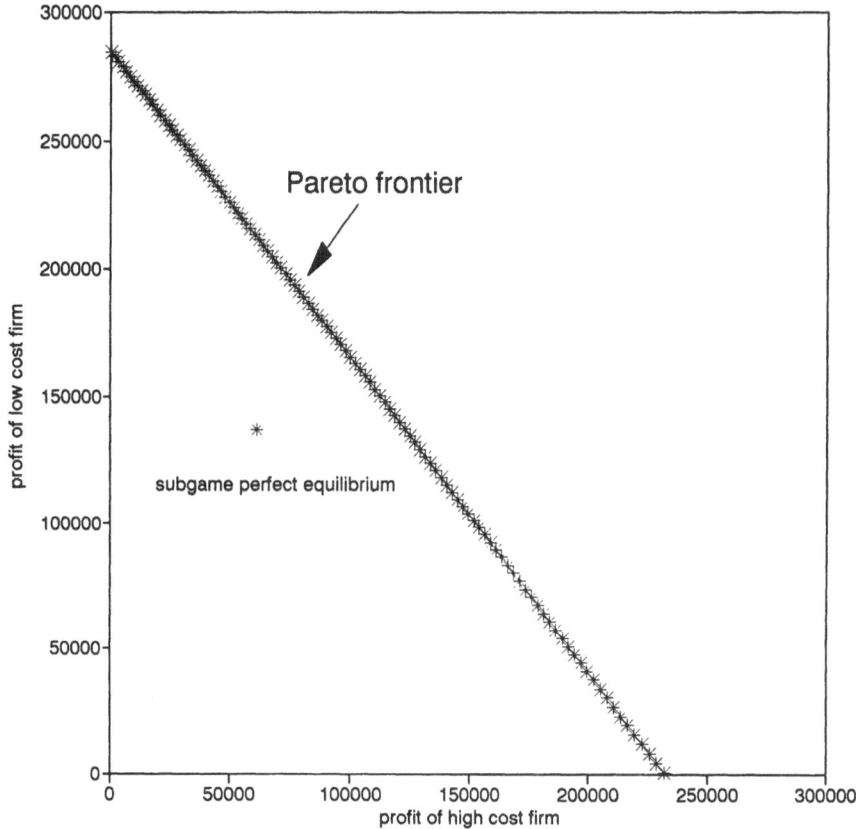

Figure 2.2: Pareto frontier and the profits in the subgame perfect equilibrium solution

2.4   A prominent cooperative solution

In the following a "prominent" cooperative solution of the game will be discussed. By prominence we mean that the solution suggests itself for the given situation. This concept of prominence goes back to Schelling (1960). As we shall see later, the prominent solution to be described seems to have an important influence on the strategic thinking of many participants of the tournament.

In the considered cooperative solution, both firms in the market always maximize their profits of the current period. They behave as if there would not be a following period. We call this the "myopic monopoly solution". The resulting prices are the "myopic monopoly prices". The myopic monopoly price of a firm i in period t is given by:

$$P_i{}^{t,M} = \tfrac{1}{2}(D_i{}^t + c_i) \tag{21}$$

Figure 2.3 shows the resulting price paths and demand potentials. We see that in the first period the high cost firm starts with a higher price than the low cost firm. It thus loses demand potential so that its next period monopoly price is lower than before. The opposite is true for the low cost firm. Thus, the myopic monopoly prices of both firms assimilate and eventually remain constant at 132. The demand potentials then remain at 207 for the low cost firm and 193 for the high cost firm. The resulting long–run profits are 157 569 for the low cost firm and 106 187 for the high cost firm. These profits lie very close to the Pareto frontier which has been computed in the previous section.

The myopic monopoly solution is not suggested by abstract economic or game theoretic concepts. However, in the specific situation of our experimental game it easily suggests itself to somebody who looks for a natural mode of cooperation as a point of departure for the construction of a strategy.

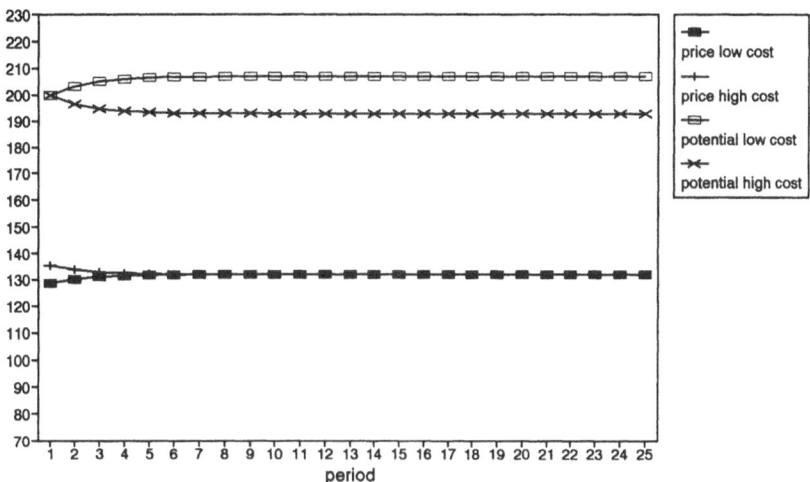

Figure 2.3: The myopic monopoly solution

3.   THE EXPERIMENTAL DESIGN

This chapter describes how the experiments were organized. I started with a series of game–playing experiments described in Section 1. Following up the game–playing experiments, I organized computer tournaments with strategies designed by subjects. The organization of the computer tournaments is explained in Section 2.

3.1   Organization of the game–playing experiments

The subjects participating in the game–playing experiments were students from the University of Bonn. They were recruited partly from an introductory economics course for first–year economics students and partly by notices on the billboards around the Department of Law and Economics. Thus, the bulk of the subjects were students of economics or of law although everybody else was admitted. Only a few of them had knowledge of oligopoly or game theory.

I organized five experimental sessions with twelve subjects each. They lasted three to four hours. The subjects made their decisions at a computer terminal and did not interact in any other way. Before the experiment started, they were instructed about the game and the use of the computer program during about three quarters of an hour. Then, they played twice the twenty–five–period game.

The twelve subjects of a session were divided into six groups of two players each so that they formed six duopoly markets. In each market, one player was a low cost player and the other was a high cost player. None of the participants knew which other person of the group was his opponent. In the second play, everybody entered the market with a different opponent but remained in the same cost situation as in the first play. More specifically, the organization of each experimental session was such that always four participants formed an independent player group. This yielded three independent player groups per session, so that in total there are 15 independent observations. This is important for statistical purposes. Figure 3.1 shows how the markets were built up out of the twelve players in the first and second play.

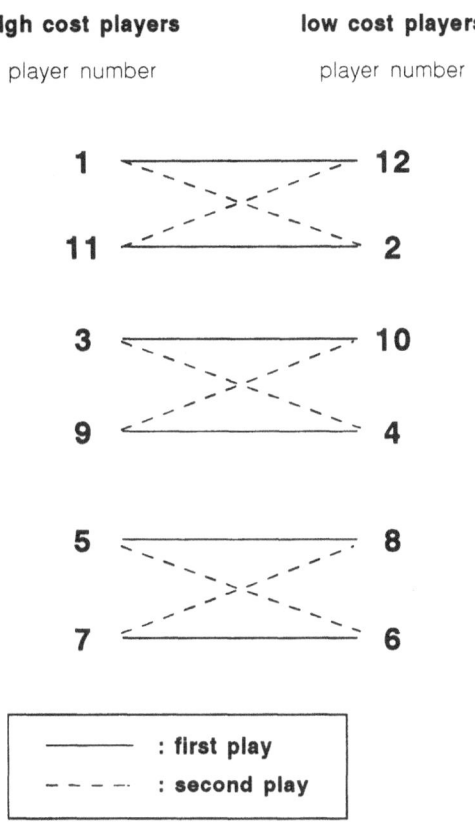

Figure 3.1: Partition of the twelve subjects of an experimental session into different markets in the two plays such that the subjects formed three independent player groups.

Each subject was sitting in a separate cubicle with a personal computer. All computers were connected through a local area network. A master program which controlled the whole experiment was installed on a supplementary computer in this network. In each of twenty–five periods, the two players in each market independently had to choose their prices which then were transmitted to the master program. When both players' prices had been chosen, the master program calculated each player's sales, profit, and new demand potential and sent all information back to both players. With this information, the players entered the next period. Both players of a market had complete information about the rules of the game and the history of their market. This information could always be called up on the screen by a selection menu. As each player in each period knew his actual demand potential he could compute his sales and profits resulting from any price he was allowed to set. He could do the same calculations also for his opponent's actual situation. The player's program also provided a calculation aid in its selection menu which reported all relevant figures for hypothetical prices to be selected by the player. If the hypothetical or final price decision of a player was out of bounds the program did not accept it and asked for a new decision.

The subjects were paid at the end of each session according to their balance after period 25 of each play. A subject's payment for a play was determined in the following way: The surplus of his total profit over the average total profit realized by all other subjects of the session who played in the same cost situation was multiplied by an exchange rate and then added to sixteen Deutsche Mark (DM).[5] Obviously, the surplus must be negative for some players if it is positive for others. The subjects knew the exchange rate and thus the equivalence of one unit of profit to payoff in DM.[6]

3.2   Organization of the strategy tournaments

I sent out about 250 information brochures (Keser (1989)) to recruit participants for the strategy tournaments which I called IDEAS – International Duopoly Experiments And Simulations. The brochure described the rules of the game. As a substitute for game–playing practice, it also gave the readers some information about the outcome of the game–playing experiments. This included protocols of some actually observed plays, showing a variety of performances. Every participant was asked to develop a strategy for each of two cost situations

---

[5] Note that this method of payment is strategically equivalent to multiplying each subject's total profit of a play with a fixed exchange rate (DM/profit point). Of these methods of payment the latter would be more straightforward but the one chosen has the advantage that the total amount of money to be paid to all subjects is fixed in advance.

[6] The exchange rate was .0002.

(low and high cost) of the multistage duopoly game with demand inertia. The strategies should be presented in flow–chart form. An instruction of how to prepare a flow–chart was offered in the information brochure. The submitted flow–charts then were translated into computer programs. The participants were told that each strategy would play in a computer simulation against each strategy for the opposite cost situation of the other participants. The objective was to reach a high profit on average over all plays. The only incentive I could give was to build up ranking lists of these profits. By announcing to publish such lists of realized profits, I hoped to prevent participants from developing "all–or–nothing" strategies, i.e. strategies designed to win the whole tournament by all means without caring about what would happen in case of failure. Although everybody could choose to play anonymously, only six participants decided to do so.

Forty–five participants from twelve different countries (according to mailing address) participated in a first tournament. Most of the participants are young economists at university institutes. After the first tournament each participant received the ranking lists of the average profits of all strategies for each cost type. On these ranking lists, each participant could only identify his own strategies and the numbers of the other participants' strategies. Additionally, he received the protocols of all the plays in which his strategies were involved.

Each participant had the opportunity to participate in a second tournament round, either with his old strategies or with a new pair of strategies. Thus, a second tournament took place with thirty–four participants. Ten of them kept their strategies unchanged.

A list of all participants and their participation numbers is given in Appendix A. Note that the participation numbers are not the ones used in the correspondence with the participants after the first tournament. Participants no. 1 to 34 are those who participated also in the second tournament round. They can identify themselves with the participation number in the second tournament. The numeration of the participants in the second tournament is such that participants no. 1 to 24 are those who have submitted modified strategy pairs while participants no. 25 to 34 have participated with the old strategies. Participants no. 35 to 45 are those who dropped out after the first tournament round.

4. RESULTS OF THE GAME–PLAYING EXPERIMENTS

In Section 1 of this chapter, we start with analyzing observed average behavior. We distinguish, however, between subjects in the role of the low cost firm and subjects in the role of the high cost firm. Furthermore, we distinguish between first and second plays. In Section 2, we look at what happened, in the second plays, on the individual markets. We concentrate on the second plays because the subjects were, by then, more familiar with the game than in the first plays.

Note that for statistical purposes, we use only standard non–parametric procedures. See for example Siegel (1957) or Lienert (1973).

4.1   General results

4.1.1 Prices and demand potentials

In Figures 4.1 and 4.2 we compare average realized prices over all low cost and high cost firms in first and second plays to equilibrium prices and myopic monopoly prices in each period. We see that the realized average prices over all low cost and all high cost firms are, except in the end, almost always above the equilibrium prices. The actual behavior in the first plays is, however, rather close to equilibrium. The graphs of the average prices in the first plays even show shapes very similar to the graphs of the equilibrium prices. It seems as if the subgame perfect equilibrium might well describe the actual behavior. However, we call this an "illusory support for the theory". The reasons will become clear when we shall look at actual individual behavior, or at the average of the realized demand potentials in each period.

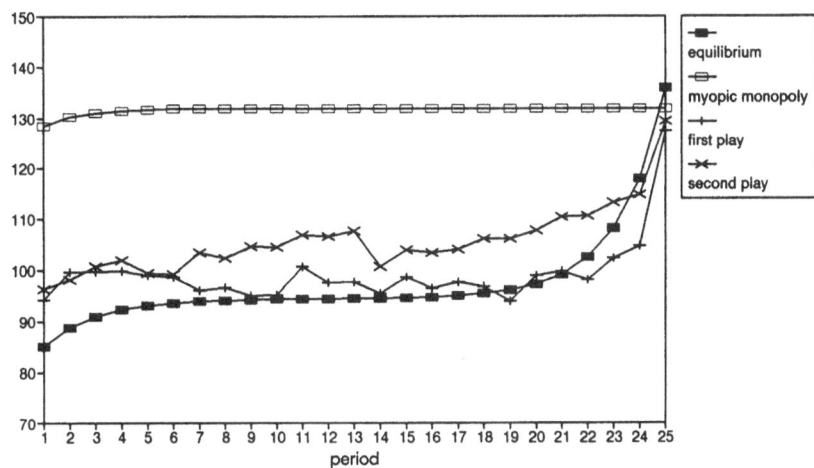

Figure 4.1: Equilibrium prices, myopic monopoly prices, and realized prices on average over all LOW COST firms

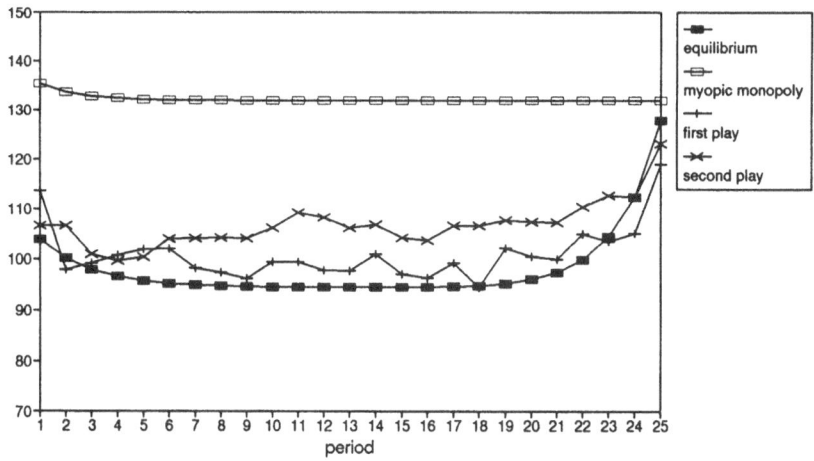

Figure 4.2: Equilibrium prices, myopic monopoly prices, and realized prices on average over all HIGH COST firms

The figures also convey the impression that there is a tendency to higher prices in second plays although they are still quite far below the myopic monopoly prices. To show that there is a significant tendency to higher prices, we compare the mean price over all 25 periods of the individual player groups in first and second plays. Table 4.1 reports the number of cases for the 15 player groups where the mean price of the low and the high cost firms and the mean market price was higher in the second play. There is a clear tendency towards higher prices in second plays. A one-tailed Wilcoxon matched pair signed rank test rejects the null hypothesis on a significance level of 5% in all cases.[7]

Table 4.1

Comparison of the mean prices of the 15 independent player groups in first and second plays

firm type	# price increases	# price decreases	α^*
low cost	11	4	5%
high cost	10	5	5%
both	11	4	5%

* Significance level of a one–tailed Wilcoxon matched pair signed rank test to reject the null hypothesis in favor of a tendency towards higher prices in second plays.

---

[7] Note that the Wilcoxon test considers not only the number of observed price increases and decreases but also the ranks of the size of the price difference between first and second play.

In Figures 4.3 and 4.4 we compare the realized demand potentials over all low cost firms and all high cost firms in first and second plays to demand potentials in equilibrium. Here, we can see that it would be wrong to conclude from Figures 4.1 and 4.2 that the equilibrium concept describes well the actual human behavior. The high cost firms give up demand potential over time but not as quickly and not as much as they would give up in equilibrium. In the second plays, they tend to give up even less demand potential than in the first plays. We may conclude that the low cost firms are not able to increase their demand potential as quickly and as much as they might do in equilibrium.

4.1.2 Profits

Table 4.2 shows the average long-run profits over all low cost and all high cost firms realized in the first and the second plays. These averages are compared to the long-run equilibrium profits and the long-run profits which result if each firm always plays its myopic monopoly price. We see that for both firm types the average long-run profit has increased in the second plays. The high cost firms have gained in both plays more than predicted by normative theory: 105% of the equilibrium profit in the first plays and 122% in the second plays. But the low cost firms reached only 83% of the equilibrium profit in the first plays and 93% in the second plays. However, if we consider the realized percentage of the profit which would be gained in the myopic monopoly solution, the low cost firms have reached a higher percentage than the high cost firms in both plays. Note that in the subgame perfect equilibrium solution the low cost firm's share in the market profit is much higher than in the myopic monopoly solution. The actual behavior of the subjects shows that the low cost firms do realize higher profits than the high cost firms but not as much higher as predicted by the subgame perfect equilibrium solution.

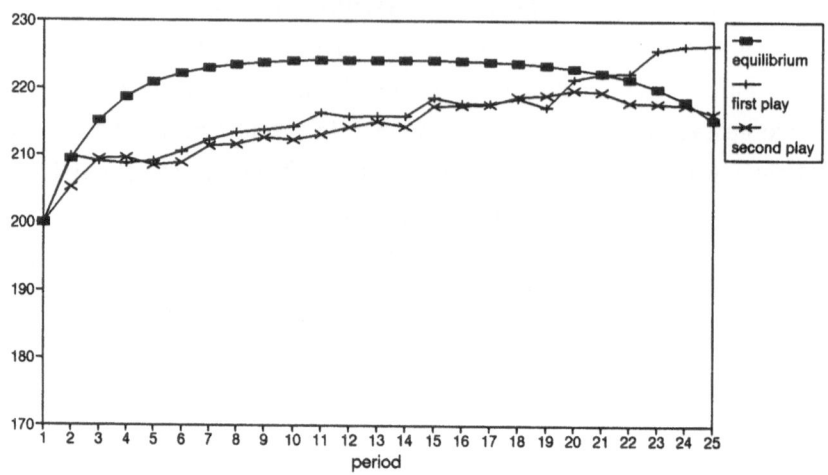

Figure 4.3:   Demand potentials in subgame perfect equilibrium and realized demand potentials on average over all LOW COST firms

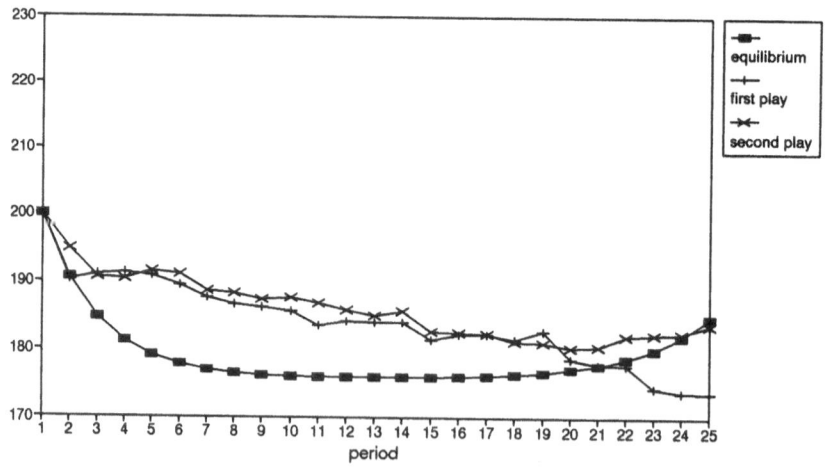

Figure 4.4:   Demand potentials in subgame perfect equilibrium and realized demand potentials on average over all HIGH COST firms

Table 4.2

Realized average long-run profits compared to long-run equilibrium profits and long-run profits in the case that in each period each firm sets its myopic monopoly price

firm type	G^*	G^m	FIRST PLAYS				SECOND PLAYS			
			av. profit	δ_{n-1}	$\%G^*$	$\%G^m$	av. profit	δ_{n-1}	$\%G^*$	$\%G^m$
low cost	136 728	157 569	114 069	27 113	83	72	127 452	23 326	93	81
high cost	61 498	106 187	64 622	34 839	105	61	75 181	31 918	122	71

G^* long-run profit in subgame perfect equilibrium

G^m long-run profit in the myopic monopoly solution

In Table 4.3 we compare the average long–run profits of the individual player groups in first and second plays. For both firm types together and for each firm type separately, we count the number of cases where the profit of the independent player groups has increased and the number of cases where it has decreased in the second play. There is a clear tendency towards higher profits in the second plays. A one–tailed Wilcoxon matched pair signed rank test rejects the null hypothesis on a significance level of 2.5% for the low cost firms and both firm types together and of 5% for the high cost firms.

Table 4.3

Comparison of the average long–run profits of the 15 independent player groups in first and second plays

firm type	# profit increases	# profit decreases	α^*
low cost	11	4	2.5%
high cost	10	5	5%
both	12	3	2.5%

* Significance level of a one–tailed Wilcoxon matched pair signed rank test to reject the null hypothesis in favor of a tendency towards higher profits in second plays.

4.1.3 Price instability

To describe a firm's price instability in an actual play we would like to have a "price instability measure". Let us define the price instability measure S_i for a firm i as :

$$S_i = \sum_{t=2}^{25} (p_i{}^t - p_i{}^{t-1})^2 \qquad\qquad i = 1, ..., 60 \qquad (22)$$

This measure computes the sum of a firm's squared price deviations from its own price in the previous period. We add up the values of this measure for all firms in each independent player group in first and in second plays and test whether the second plays are less instable. In 12 out of 15 independent player groups we observe a lower price instability in the second plays. A one–tailed Wilcoxon matched pair signed rank test rejects the null hypothesis in favor of a decrease of price instability on a significance level of 5%.

4.2   Analysis of the individual markets of the second plays

Building averages over all markets is not an appropriate way of analyzing the data if one's aim is to find out typical characteristics of behavior. We have to analyze the behavior on the individual markets. However, the observed market behavior varies considerably. Thus, our next step is to determine different types of actual market behavior. We do this first on the basis of visual inspection of the graphs of the price dynamics in each market.[8] This inspection leads us to distinguish between three main types of behavior.

---

[8] Visual inspection means that we tried to describe the observed behavior in each market and then compared markets in order to find groups of markets with similar performance.

First, there are "strongly cooperative" markets where both firms permanently set relatively high prices. Either they start already setting prices on the same level or they start with different prices in the first periods and then "agree" on a price from which they deviate only slightly over time such that there are only moderate price changes. Thus, demand potentials either do not differ very much or, if there are price differences in the beginning, they remain relatively constant in favor of the firm with the lower initial prices. Profits are very high for both firms, although given the different cost situations of the two firms, the profits of the low cost firms are in most cases considerably higher than those of the high cost firms.

Secondly, there are "aggressive" markets. The most typical markets in this category are those where the low cost firm tries to force the high cost firm out of the market by setting extremely low prices. The low cost firm exploits the advantage of still making profits at a price level just above or at the unit production cost of the high cost firm. It can even go below the opponent's unit cost such that it takes away demand potential from the high cost firm in any case. A high cost firm has hardly any chance against such an opponent: it cannot make reasonable profits at the prevailing price level but it also cannot increase its own price without having its demand potential reduced. Typically, a high cost firm facing such an aggressive opponent defends itself by permanently switching from relatively high to low prices in order to make some profit in some periods and to improve demand potential in others. Both firms end up with very low profits.

Thirdly, there are cases between the aggressive and the strongly cooperative ones, the "weakly cooperative" markets. The average prices are neither extremely high nor very low but there is a permanent movement in prices. Typically, there are at some times abrupt price cuttings by a firm which are not expected by the other. The possible idea behind these price cuttings is to first increase one's demand potential such that in the following period the price can be increased above the initial price level. The firm makes a relatively high profit in this following period without ending up with a very low demand potential. (Indeed, some subjects told us after the experiments that they behaved in such a way in order to make "super profits" from time to time.) Sometimes both firms in a market show such a behavior, sometimes there is only one firm which behaves occasionally according to this idea while the other one either follows the price variations of the opponent firm or just remains at the initial price level. Typically, advantages in demand potential shift several times from one firm to the other. But often, the high cost firm after a while remains with a lower demand potential than the low cost firm. Both firms make long–run profits which are neither very low nor very high. The long–run profit of the low cost firm is always higher than the profit of the high cost firm.

All observed markets are captured by this clustering. Among 30 markets, we identify 9 strongly cooperative, 10 weakly cooperative, and 11 aggressive markets. An example of each market type is graphically represented in Appendix B.

Our next step in characterizing markets is to identify some measures which allow us to distinguish between the three market types by numerical classifications. We consider

- market profit: the sum of the long–run profits of the two firms in a market
- price level: the average price of the two firms in a market and over the 25 periods
- price instability: the price instability measure defined in Chapter 4.1.3, added up for the two firms in a market

In Figure 4.5, a plot of the price levels and price instability measures of the markets, we again can distinguish between the three types described above. We draw the convex hull of all the observations in each category. There is a clear gap in the price level which allows us to separate aggressive and cooperative markets. The distinction between strongly and weakly cooperative markets is less obvious. The strongly cooperative markets show less price instability than the weakly cooperative ones. Thus, a combination of high prices and a low price instability characterizes the strongly cooperative markets while the weakly cooperative markets show sometimes relatively high prices but always a high price instability. Surprisingly, this clustering, which is based only on mean values over all periods, is fully in concordance with the ad hoc distinction described above, which also takes the price dynamics into account.

As average prices and long–run profits of the observed markets are highly correlated, Figure 4.6, which is a plot of the market profits and price instability measures of the markets, is very similar to Figure 4.5 and proposes the same clustering of the markets. Firms in cooperative markets make clearly higher profits than those in aggressive ones. Again, strongly and weakly cooperative markets are to be distinguished by their price instability.

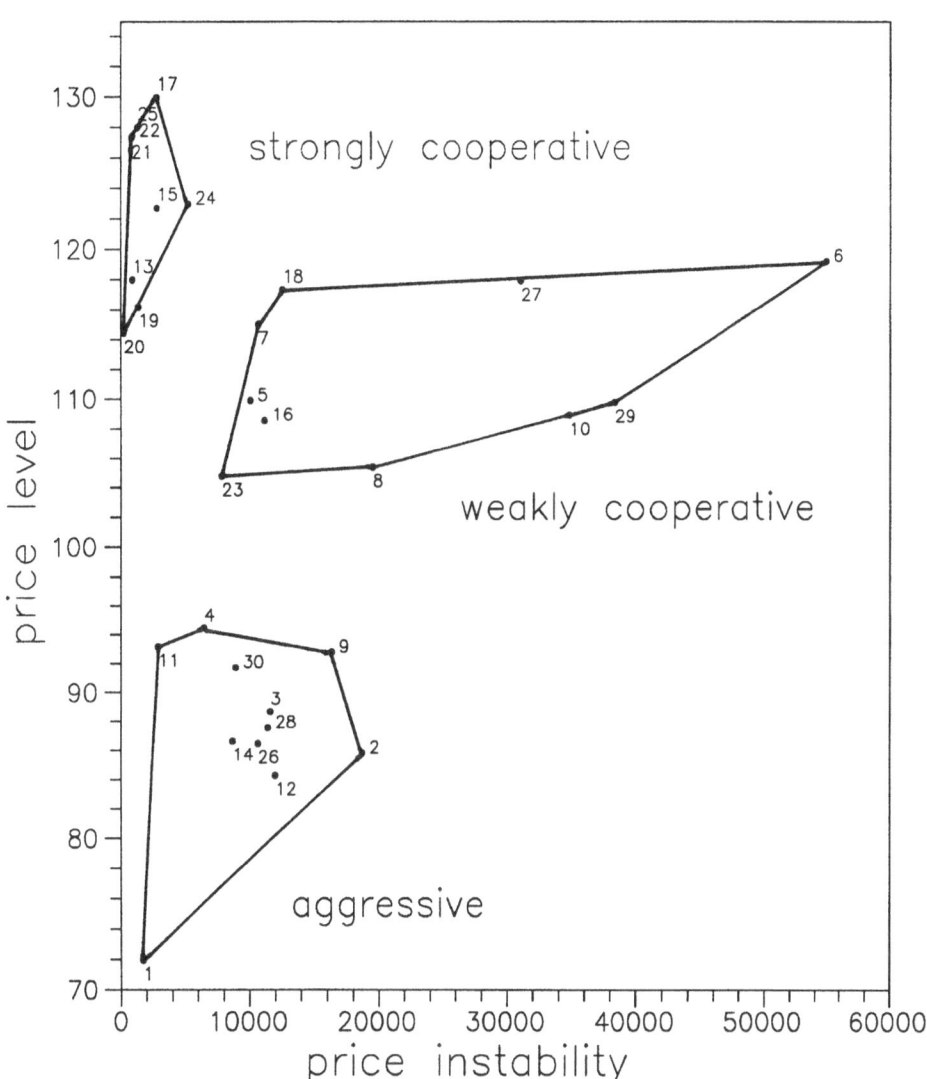

Figure 4.5:   Average prices and price instability of the markets

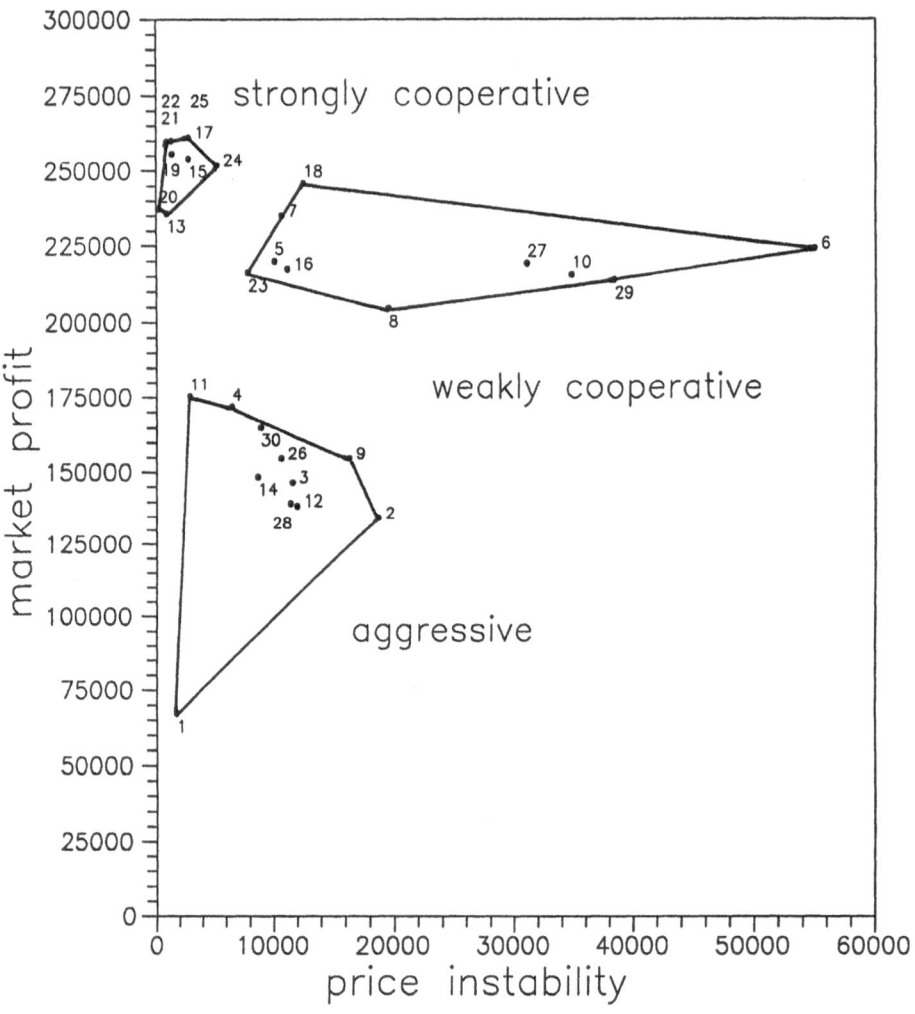

Figure 4.6: Market profits and price instability of the markets

Table 4.4 yields for each market type the average market profit, price level and price instability.

Table 4.4

Averages of market profits, price levels and price instability over all markets of each type

measure	market type		
	aggressive	weakly cooperative	strongly cooperative
# markets	11	10	9
market profit	145 078	221 151	252 624
price level	88	112	123
price instab.	9 928	23 101	1 807

Recall that we have clustered the markets according to the following criteria:

- Cooperative markets show on average over all periods higher prices than aggressive markets
- Cooperative markets show on average over all periods higher profits than aggressive markets
- Strongly cooperative markets are less price instable than weakly cooperative markets

Furthermore, Table 4.4 leads us to the following hypotheses:

- The average prices on strongly cooperative markets are higher than the prices on weakly cooperative markets (H_1)
- Strongly cooperative markets make higher profits than weakly cooperative markets (H_2)
- Strongly cooperative markets are less price instable than aggressive markets (H_3) and aggressive markets are less price instable than weakly cooperative markets (H_4)

One-tailed U-tests against H_1, H_2, H_3, and H_4 reject the respective null hypotheses that such differences do not exist on the following significance levels:

$$H_1 : \quad .1\%$$
$$H_2 : \quad .1\%$$
$$H_3 : \quad .1\%$$
$$H_4 : \quad 2.5\%$$

Figure 4.7 is a plot of the long-run profits of the high cost firms and the long-run profits of the low cost firms. The realizations of the actual markets are indicated by *dots* while the *stars* repesent the points of the Pareto frontier. We see that some of the strongly cooperative markets make profits very close to the Pareto frontier. Aggressive markets can be clearly distinguished from cooperative markets by their distance to the Pareto frontier. We also see that the long-run profit of the high cost firms is a very strong indicator to identify aggressive and cooperative markets. The high cost firms of the aggressive markets make a lower profit than the high cost firms of the cooperative markets. But the profit of the low cost firm alone is not reliable an indicator for the type of market.

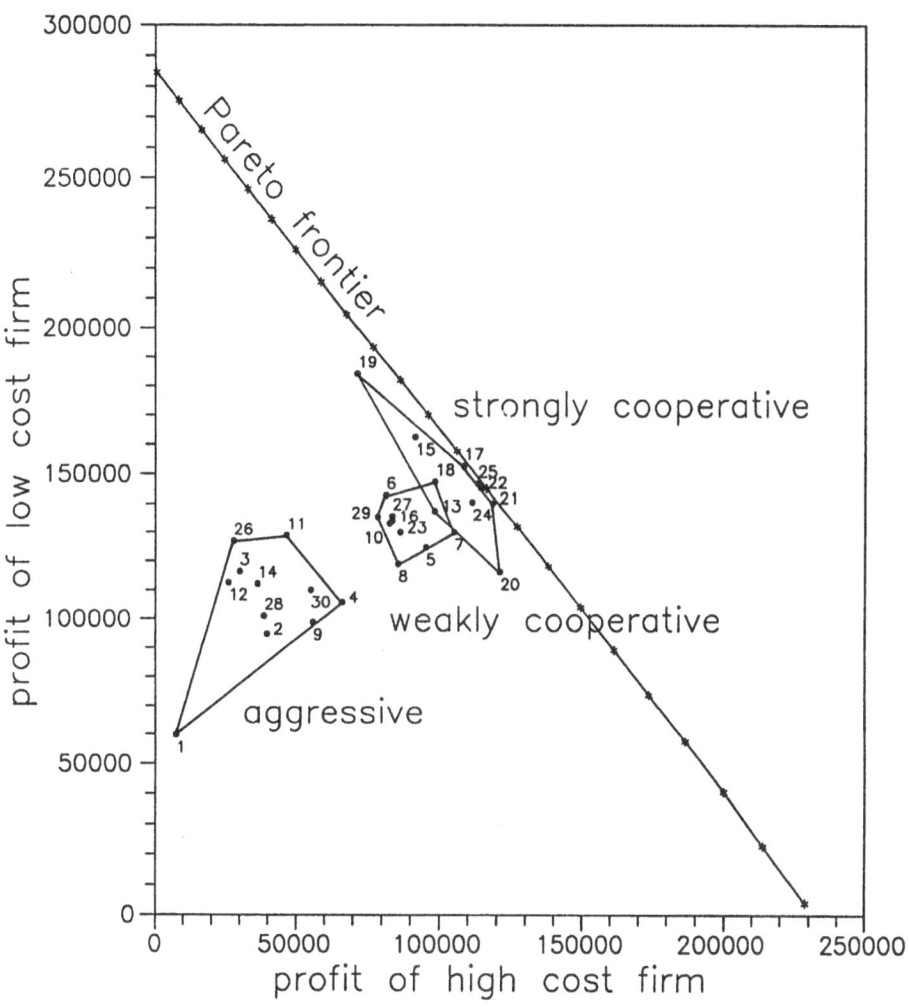

Figure 4.7: Realized long–run profits of low cost and high cost firms and Pareto frontier

5. RESULTS OF THE STRATEGY TOURNAMENTS

5.1 General Results

In the computer tournaments, the success of a strategy is measured by its realized average profit over all plays. Therefore, we will analyze the tournaments first with respect to realized profits. Let me mention here that two participants submitted for both tournaments strategies with random decisions. To reduce the effect of the randomness, these strategies, in each tournament, are played nine times against every other strategy and we consider the average profits realized over the nine plays.[9]

5.1.1 Average profits in the tournaments

For a rough comparison of results in game–playing experiments and the two tournaments, we start by looking at the realized average profits over all subjects of each firm type in the game–playing experiments and the realized average profits over all strategies for each firm type in the tournaments. Table 5.1 shows the realized average profits and, in brackets below, the correspondig standard deviations. It also shows the long–run profits that would be gained in subgame perfect equilibrium and in the myopic monopoly solution.

Due to the large cost difference, the low cost firms' profit is always higher than the high cost firms' profit. The profits in subgame perfect equilibrium are much lower than in the myopic monopoly solution for both firm types.

---

[9] I view the ninefold repetition as sufficient since the influence of the randomness is not very large. In the strategies of one participant, a random price is used to initiate a search process and, in the strategies of the other participant, the price decision is disturbed by a small random influence.

Table 5.1

Realized average profits in the game-playing experiments and the strategy tournaments compared to the profits in subgame perfect equilibrium and in the myopic monopoly solution

firm type	G^*	G^m	average profit (standard deviation) realized in			
			game-playing experiments (30 markets)		strategy tournaments	
			first play	second play	first round (45 strategies)	second round (34 strategies)
low cost	136 728	157 569	114 069 (27 113)	127 452 (23 326)	140 920 (24 324)α	140 406 (6 644)α
high cost	61 498	106 187	64 622 (34 839)	75 181 (31 918)	90 447 (10 362)α	90 206 (10 663)α

* G^* profit in subgame perfect equilibrium

G^m profit in the myopic monopoly solution

α   standard deviation of the strategies' average profit over all plays

In the game–playing experiments we see that, although average profits increased from the first to the second plays for both firm types, they were still much lower than the profits in the myopic monopoly solution. In both plays, the high cost firms obtained average profits that were higher than in subgame perfect equilibrium, while the low cost firms on average fell short of the subgame perfect equilibrium profit.

In each strategy tournament, both firm types realize on average higher profits than in the game–playing experiments and higher profits than in subgame perfect equilibrium. Average profits in the second tournament are almost the same as in the first. We must, however, be careful with these averages over all firms since the standard deviations in most cases are very high. For example, the average profits are not a good measure with which to compare the success of the low cost strategies in the first and the second tournament since the corresponding standard deviations are very different. Furthermore, the population of participants is different in the two tournaments.

5.1.2   Ranking lists of the strategies' average profits

The participants were told that the aim should be to reach with each strategy as high a profit as possible on average over all plays against the other participants' strategies of the other cost type. In order to provide them with an incentive to do well, I announced that ranking lists of these profits would be published. We are, however, not so much interested in the rankings as such but in the level of realized profits. Tables 5.2 and 5.3 show the ranking lists of the average profits in both tournaments, separately for each firm type. In these lists we also find indicated where the myopic monopoly profits, the equilibrium profits, and the realized average profits would have been in the ranking.

We see that most of the profits in a ranking list do not differ very much from each other. Only slight changes in profits would be enough to change positions. These rankings might, therefore, easily change, for example with the addition of a new strategy.

Table 5.2

Average profits of the LOW COST strategies in the

first tournament   second tournament

rank	average profit		strategy no.
1	159 972		29
2	158 349		9
	157 569	G_m	
3	156 426		33
4	156 052		11
5	155 865		17
6	155 682		43
7	155 432		25
8	154 573		41
9	152 584		4
10	151 551		13
11	151 068		10
12	150 368		31
13	150 118		20
14	149 726		23
15	149 608		42
16	149 522		5
17	149 289		18
18	148 727		6
19	148 589		40
20	148 399		34
21	148 319		8
22	148 253		24
23	148 214		39
24	148 111		26
25	147 834		30
26	147 833		35
27	147 825		27
28	147 232		14
29	145 981		1
30	145 733		28
31	145 688		22
32	144 640		32
33	143 493		15
34	140 992		38
	140 920	∅	
35	140 567		19
36	140 088		37
37	137 977		44
	136 728	G*	
38	136 595		36
39	135 982		21
40	135 956		2
41	129 627		7
42	99 334		12
43	76 941		16
44	76 931		3
45	29 340		45

rank	average profit		strategy no.
	157 569	G_m	
1	148 485		10
2	147 949		9
3	147 543		18
4	147 028		17
5	146 661		11
6	146 454		29
7	145 869		22
8	145 514		13
9	145 120		31
10	144 829		23
11	144 622		8
12	144 115		24
13	144 085		5
14	144 026		20
15	141 995		30
16	141 890		26
17	141 308		34
18	141 152		14
19	140 406		6
	140 406	∅	
20	140 396		27
21	140 395		1
22	140 307		3
23	140 284		32
24	139 335		25
25	138 269		33
26	138 120		28
27	138 108		16
	136 728	G*	
28	135 157		12
29	135 094		4
30	133 846		19
31	131 385		2
32	131 142		7
33	122 432		15
34	120 489		21

G* :   profit in subgame perfect equilibrium
G_m :   profit in the myopic monopoly solution
∅ :   average over all strategies

Table 5.3

Average profits of the HIGH COST strategies in the

first	tournament			second	tournament		
rank	average profit		strategy no.	rank	average profit		strategy no.
	106 187 G$_m$				*106 187* G$_m$		
1	100 953		42	1	99 581		29
2	99 961		29	2	99 013		22
3	98 327		25	3	98 930		10
4	98 026		43	4	98 057		9
5	96 861		17	5	97 214		11
6	96 583		10	6	97 059		17
7	95 926		31	7	95 609		18
8	95 722		33	8	95 437		15
9	95 440		8	9	94 636		5
10	95 371		39	10	94 589		8
11	95 061		15	11	94 559		25
12	94 693		5	12	94 442		31
13	94 682		3	13	93 910		23
14	94 662		16	14	93 736		3
15	94 312		44	15	93 599		30
16	94 217		23	16	93 348		13
17	93 868		11	17	93 065		7
18	93 711		34	18	92 974		6
19	93 583		14	19	92 695		14
20	93 336		18	20	92 255		33
21	93 247		7	21	90 631		27
22	93 140		13		*90 206*	Ø	
23	92 908		30	22	90 180		20
24	92 883		19	23	89 514		28
25	92 758		27	24	89 511		26
26	92 694		38	25	89 508		16
27	92 571		26	26	89 479		34
28	92 523		28	27	88 908		4
29	92 497		1	28	86 620		1
30	92 497		22	29	85 447		19
31	92 491		37	30	84 750		32
32	91 870		9	31	84 694		24
33	91 709		20	32	80 595		2
34	91 399		40	33	74 213		12
35	91 070		4		*61 498*	G*	
	90 447	Ø		34	38 251		21
36	90 265		41				
37	88 006		32				
38	86 275		24				
39	85 892		36				
40	84 603		35				
41	83 918		6				
42	79 367		2				
43	65 384		21				
44	63 571		12				
	61 498	G*					
45	41 299		45				

G* : profit in subgame perfect equilibrium
G$_m$: profit in the myopic monopoly solution
Ø : average over all strategies

For the low cost strategies, the range of profits in the first tournament is very different from that of the second tournament. In the first tournament, we observe extremely high profits as well as extremely low ones. Two strategies realize profits even higher than the profit which could be gained in the myopic monopoly solution. The lowest profits are even more extreme and affect the average to the extent that three quarters of the strategies realize profits above the average. Such outliers do not occur in the second tournament. Although the average profits are almost the same in the two tournaments, the median is somewhat lower in the second tournament. The winning strategy of the second tournament would have been only twentieth in the first one.

The ranking lists for the high cost strategies seem very much alike for both tournaments. However, there is a problem with these lists when we would like to analyze the success of a certain type of strategy in the considered tournament.[10] Since a participant's strategy is only played against the strategies of all the other participants, each participant's strategy plays against a different population. It turns out to be important that the participants' own counterparts are excluded. Thus, for example in the first tournament, a high cost strategy always fixing its myopic monopoly price does relatively well in case the low cost strategy of the same participant always sets a price of 71 (strategies 3 and 16), while it does less well in case the corresponding low cost strategy also sets its short–run monopoly price in each period (strategy 28). In the first case, the excluded strategy is aggressive while in the second case it is cooperative.

For this reason we repeat the tournaments in such a way that each strategy also plays against its own counterpart. We use the resulting rankings, which are given in Tables 5.4 and 5.5, for analytical purposes in Chapters 5.2 and 5.3. They are, indeed, not so different from the original ones for the low cost strategies. Only in the high cost rankings, there occur some drastic changes of profit and position. The average profits over all strategies, which are given in Table 5.6, increase slightly in the new tournaments, presumably because many strategies cooperate with their counterparts.

---

[10] See Chapters 5.2 and 5.3 below.

Table 5.4

Average profits of the LOW COST strategies in modified tournaments
where each strategy plays also against its counterpart

| | first tournament | | | second tournament | |
rank	average profit	strategy no.	rank	average profit	strategy no.
1	160 211	29	1	148 633	10
2	158 243	9	2	147 947	9
3	155 956	33	3	147 695	18
4	155 867	43	4	147 339	17
5	155 756	17	5	147 168	29
6	155 508	11	6	146 842	22
7	155 016	25	7	146 381	11
8	154 354	41	8	145 875	13
9	152 780	4	9	145 480	31
10	151 685	13	10	145 204	23
11	151 221	10	11	144 939	8
12	150 524	31	12	144 476	5
13	149 997	20	13	144 374	20
14	149 900	23	14	143 954	24
15	149 603	5	15	142 366	30
16	149 491	42	16	142 351	26
17	149 470	18	17	141 660	14
18	148 960	40	18	141 431	34
19	148 447	8	19	140 864	6
20	148 334	34	20	140 815	3
21	148 321	26	21	140 775	27
22	148 282	39	22	140 394	32
23	148 172	24	23	139 698	1
24	148 079	6	24	139 258	25
25	147 986	35	25	138 692	28
26	147 985	30	26	138 680	16
27	147 946	27	27	138 180	33
28	147 481	14	28	136 034	4
29	146 239	1	29	135 817	12
30	145 996	28	30	131 033	19
31	145 952	22	31	130 876	2
32	144 626	32	32	130 674	7
33	143 323	15	33	123 830	15
34	141 346	38	34	118 946	21
35	140 937	19			
36	140 176	37			
37	138 203	44			
38	135 874	2			
39	135 478	21			
40	134 747	36			
41	128 210	7			
42	100 634	12			
43	77 300	16			
44	77 290	3			
45	29 881	45			

Table 5.5

Average profits of the HIGH COST strategies in modified tournaments
where each strategy plays also against its counterpart

first tournament			second tournament		
rank	average profit	strategy no.	rank	average profit	strategy no.
1	101 280	42	1	98 999	29
2	99 513	29	2	98 993	10
3	97 983	43	3	98 124	9
4	97 508	25	4	98 043	22
5	96 777	10	5	96 632	17
6	96 389	17	6	96 459	11
7	96 146	31	7	95 783	18
8	95 591	8	8	94 953	5
9	95 473	39	9	94 858	8
10	95 451	33	10	94 777	31
11	94 839	5	11	94 583	15
12	94 739	44	12	94 271	23
13	94 483	23	13	94 101	3
14	94 208	15	14	94 096	7
15	94 193	34	15	94 038	30
16	93 905	7	16	93 717	13
17	93 826	14	17	93 587	25
18	93 625	18	18	93 393	6
19	93 504	11	19	93 042	14
20	93 430	13	20	91 999	33
21	93 255	30	21	91 028	27
22	93 185	19	22	90 690	20
23	93 011	27	23	90 240	34
24	93 005	38	24	90 004	28
25	92 977	37	25	90 001	26
26	92 873	26	26	89 998	16
27	92 843	3	27	88 409	4
28	92 826	28	28	85 904	1
29	92 823	16	29	85 687	32
30	92 801	1	30	85 208	24
31	92 801	22	31	83 874	19
32	92 241	20	32	80 869	2
33	91 808	9	33	75 154	12
34	91 585	40	34	38 289	21
35	90 741	4			
36	90 280	41			
37	88 641	32			
38	86 863	24			
39	85 083	35			
40	84 783	36			
41	84 028	6			
42	79 526	2			
43	64 980	21			
44	64 518	12			
45	42 117	45			

Table 5.6

Average profits in the modified first and second tournament
where each strategy also plays against its counterpart

firm type	average profit in	
	first tournament	second tournament
low cost	140 929	140 549
high cost	90 499	90 288

Since the average profit over all plays in a tournament increases when we repeat the tournament in such a way that each strategy also plays against its own counterpart, we suppose that profits against counterparts generally are higher than the average profits realized in the plays against all others. Therefore let us consider Table 5.7. This table reports, for the first and the second tournament, the realized profits of the two strategies submitted by a participant when playing against each other.

Table 5.7

Profit of each strategy in the play against its counterpart

participation no.	first tournament		second tournament	
	profit of low cost strategy	profit of high cost strategy	profit of low cost strategy	profit of high cost strategy
1	157 569 *	106 187 *	116 693	62 280
2	132 307	86 530 *	114 076	89 915 *
3	93 112 *	11 922	157 569 *	106 138 *
4	161 390 *	76 260	167 048 *	71 954
5	153 132 *	101 258 *	157 409 *	105 430 *
6	119 585	88 869 *	155 969 *	107 206 *
7	65 861	122 840 *	115 229	128 135 *
8	154 041 *	102 243 *	155 404 *	103 755 *
9	153 589	89 098	147 866	100 332 *
10	157 941 *	105 302 *	153 503 *	101 049 *
11	131 541	77 499	137 151	71 537
12	157 569 *	106 187 *	157 569 *	106 187 *
13	157 569 *	106 187 *	157 800 *	105 899 *
14	158 433 *	104 498 *	158 433 *	104 498 *
15	135 847	56 682	169 974 *	66 400
16	93 112 *	11 922	157 569 *	106 187 *
17	150 957	75 612	157 607 *	82 543
18	157 415 *	106 312 *	152 705 *	101 533 *
19	157 248 *	106 444 *	38 224	31 950
20	144 680	115 640 *	155 853 *	107 515 *
21	113 292	47 212	68 040	39 546 *
22	157 569 *	106 187 *	178 951 *	66 039
23	157 569 *	106 187 *	157 569 *	106 187 *
24	144 582	112 754 *	138 663	102 173 *
25	136 728	61 489	136 728	61 498
26	157 569 *	106 187 *	157 569 *	106 187 *
27	153 250 *	104 134 *	153 250 *	104 134 *
28	157 569 *	106 187 *	157 569 *	106 187 *
29	170 728 *	79 802	170 728 *	79 802
30	154 603 *	108 521 *	154 603 *	108 521 *
31	157 355 *	105 839 *	157 355 *	105 839 *
32	144 028	116 591 *	144 028 *	116 591 *
33	135 246	83 526	135 246	83 526
34	145 478	115 366 *	145 478 *	115 366 *
35	154 744 *	106 176 *		
36	53 456	35 977		
37	144 040 *	114 385 *		
38	156 930 *	106 672 *		
39	151 293 *	99 988 *		
40	165 317 *	99 758 *		
41	144 707	90 917 *		
42	144 358	115 635 *		
43	164 006 *	96 104		
44	148 152 *	113 554 *		
45	53 712 *	78 106 *		

*: higher than the profit realized on average over all plays against all other participants' strategies

We are interested in how many strategies make higher profits when playing against their own counterparts than the profits realized on average against the strategies for the other cost type of all other participants. These strategies are characterized in Table 5.7 by an *asterix*. In the first tournament (with 45 participating strategies for each cost type), we count 32 high cost strategies and 27 low cost strategies marked by an asterix. In the second tournament (with 34 participating strategies for each cost type), there are on each cost side 24 strategies whose profits against their counterparts are higher than their average profits against the opponent strategies of all other participants. Let us apply a one–tailed sign test for our hypothesis that strategies, in the play against their own counterparts, are likely to make profits higher than their average profits against the opponent strategies of the other participants. On the 2.5% significance level we reject, for the high cost strategies in both tournaments and for the low cost strategies in the second tournament, the null hypothesis of equal probability for making a higher or a lower profit against one's counterpart. For the low cost strategies in the first tournament, we cannot reject the null hypothesis on a significance level of 10%.

The result of this test gives us at the same time evidence about the statistical significance of the observed increases of the average profits over all strategies when we repeat the tournaments including plays against the own counterpart. We may, on the same significance levels, reject the null hypotheses of an equal probability of a profit increase and a decrease in favor of a profit increase for the high cost strategies in both tournaments and the low cost strategies in the second. We may not reject, on a significance level of 10%, the null hypothesis for the low cost strategies of the first tournament.

A rank correlation analysis between the strategies' profits against their counterparts and their average profits against all other participants' strategies for the other cost type yields that, for no strategy type in no tournament, there is a significant correlation. So, we can neither conclude that strategies making higher profits in the play against their counterparts do, on average against all others, better than strategies making less profits in the play against their counterparts (positive correlation), nor that they do worse (negative correlation).

5.1.3 Profits in individual markets

We would like to have some information about the individual plays in the tournaments. We are particularily interested in the net total profit of the low cost firm and the high cost firm in each market (i.e. play). It is, however, difficult to show these profits in a diagram with low cost and high cost firms' profits on the axes because there are too many markets: With 45 participating strategies for each cost type in the first tournament and 34 strategies in the second tournament we can observe 45 x 45 = 2015 plays in the first tournament and 34 x 34 = 1156 plays in the second one. Therefore, we count the frequency of observations in given profit intervals. Furthermore, for better comparison between first and second tournament, we compute the percentage of observed markets falling into the considered field. These percentages are represented in Figures 5.1 and 5.2. A dot means that some observations fall into the considered field, but less than one percent of all the observations.

In the first tournament, 45% of the observations lie in the field of profit between 140 000 and 160 000 for the low cost firm and between 100 000 and 120 000 for the high cost firm. We may consider this the "field of cooperative behavior". In the case that each firm always sets its myopic monopoly price, the resulting profits lie in these ranges. Seven percent of the markets even play exactly the myopic monopoly solution. In the second tournament, 42% of the observations are in the field of cooperative behavior. The observations are a little less spread but otherwise Figure 5.2 looks very similar to Figure 5.1. Four percent of the markets play exactly the myopic monopoly solution.

50

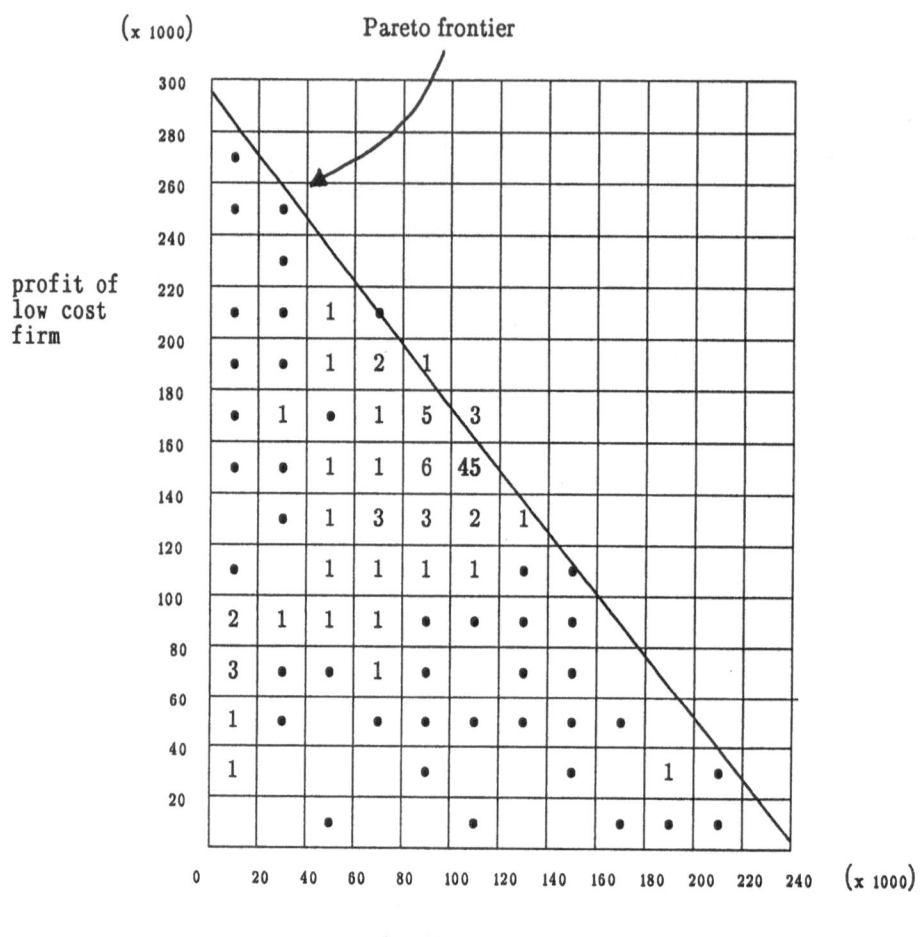

Figure 5.1:   Profits of the two firms in each market of the first tournament.
The numbers represent percent of the 2015 plays (each strategy
playing against each strategy of the other cost type) falling
into the considered field of profit for low cost and high cost
firm. A dot means that some plays fall into the considered profit
field but less than one percent of all plays.

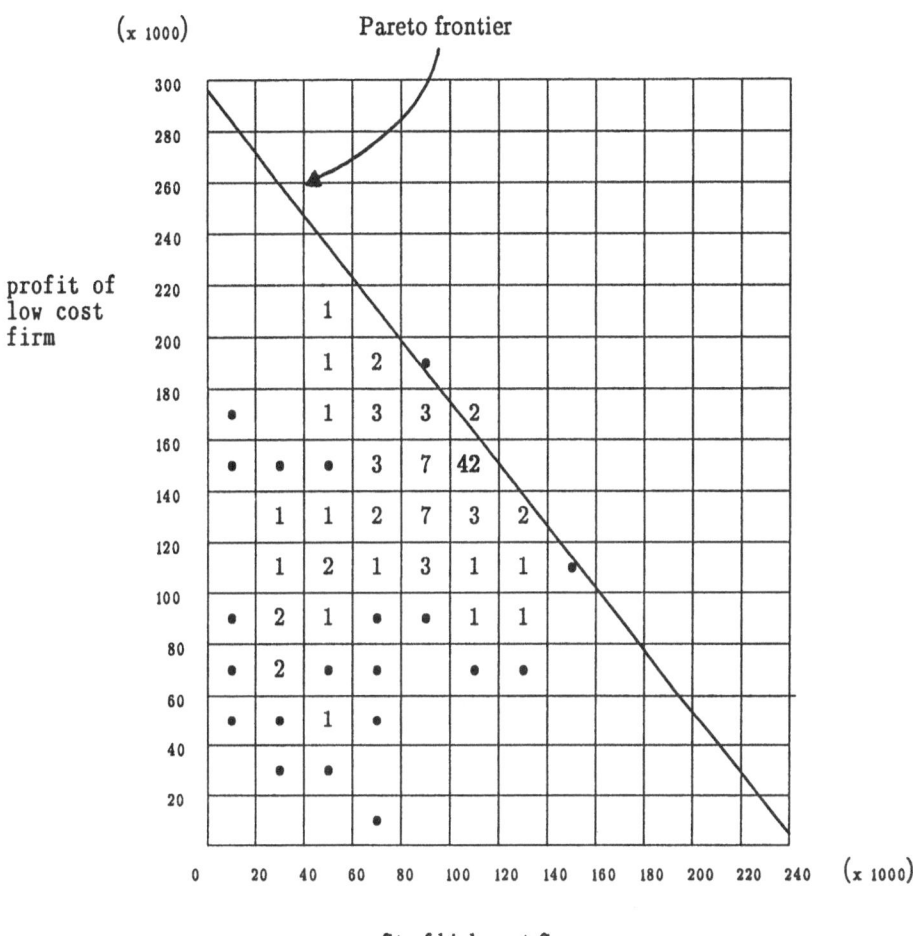

Figure 5.2: Profits of the two firms in each market of the second tournament. The numbers represent percent of the 1156 plays (each strategy playing against each strategy of the other cost type) falling into the considered field of profit for low cost and high cost firm. A dot means that some plays fall into the considered profit field but less than one percent of all plays.

5.1.4 Aggressiveness of the strategies

We have got the impression that, in both tournaments, the winning strategies are to some extent more aggressive or a little bit less cooperative than the majority of the other strategies, although not very aggressive. Some rather aggressive strategies do very badly. Therefore, we would like to have a measure for the aggressiveness of a strategy. We shall look at the average profit of all opponents playing against a given strategy as a measure of the aggressiveness of this strategy.

Our assumption is that the most successful and unsuccessful strategies are more aggressive than the the strategies with medium success, in the sense that their opponents do not make profits as high as opponents do against the strategies with medium success. To test the statistical significance, let us apply a quartile–median test. We group our strategies according to their average long–run profits in the tournament in profit "quartiles" Q1 to Q4 such that Q1 contains the most successful and Q4 the least successful strategies. The hypothesis we want to test is that the dispersion of profits of the more aggressive strategies is larger than the dispersion of profits of the less aggressive strategies. This means that the profits of the less aggressive strategies are more likely to lie in the quartiles Q2 and Q3 while the more aggressive strategies are more likely to make profits either in the quartiles Q1 of highest profits or Q4 of lowest profits. Therefore, we take the conjunction of Q2 and Q3 as "inner quartiles" and Q1 and Q4 as "outer quartiles" into consideration. Note that the number of participating strategies is in none of the tournaments dividable by four. We handle this problem by making the inner quartiles larger than the outer ones. We compute the frequency of strategies with opponents' profit below the median (i.e. more aggressive strategies) and strategies with opponents' profits above the median (i.e. less aggressive strategies) in the inner and outer quartiles so that we can write the 2x2–tables for each cost type in each tournament represented in Table 5.8. A one–tailed χ^2 test applied to each of them allows us to reject the null hypothesis of no difference of dispersion of profits of the more and the less aggressive strategies on a 5% significance level. Note that we have considered here the profits and opponents' profits of the official tournament but we get the same statistical result if we consider the tournament variation where each strategy also plays against its own counterpart.

Table 5.8

Success and aggressivity of the low cost and the high cost strategies in both
tournaments

Low cost strategies in first tournament

aggressiveness

	opponents' profit ≤ median	> median
outer profit quartiles	16	6
inner profit quartiles	7	16

success

Low cost strategies in second tournament

aggressiveness

	opponents' profit ≤ median	> median
outer profit quartiles	11	5
inner profit quartiles	6	12

success

High cost strategies in first tournament

aggressiveness

	opponents' profit ≤ median	> median
outer profit quartiles	15	7
inner profit quartiles	8	15

success

High cost strategies in second tournament

aggressiveness

	opponents' profit ≤ median	> median
outer profit quartiles	11	5
inner profit quartiles	6	12

success

Before we turn to the analysis of the structure of the strategies in the following chapter, let us summarize the main results of this chapter:

- Cooperative behavior occurs in many plays of both tournament rounds.
- The most successful as well as the most unsuccessful strategies in both tournaments are more aggressive than the strategies with medium success.

5.2   Structure of the strategies

Our main point of interest in organizing the strategy tournaments is to get insight into the structure of the strategies used by boundedly rational agents when playing the considered game. The submitted strategies show a great variety of decision rules but we are able to classify them roughly and to recognize a typical pattern of behavior which occurs very often and which explains how the observed cooperation comes into being. An important feature of the strategies is a phase structure which we shall analyze before going into the classification of the strategies.

5.2.1   Phase structure

We can observe a phase structure in most of the strategies: They check whether they are in the beginning, the middle part, or the end of the game and then behave according to different rules.

An initial phase is necessary for most strategies since they make, in the main part of the game, their decisions dependent on observations of the previous period (or periods) and thus need a specified starting price. There are, however, only few strategies which use an initial phase of more than one period. The longest initial phase which occurs is seven periods long. The intention behind the use of an initial phase of more than one period might be, in most strategies, to build up with aggressive initial prices a high market share which should guarantee high profits in the main phase of the game. Other strategies, however, set prices to be considered as cooperative during the whole initial phase, possibly to signal a willingness for cooperation. Figure 5.3 shows the percentages with which no initial phases, phases of one period and longer initial phases occur in each tournament for each cost type. The prices set in the first period will be analyzed with respect to the strategies' classification in Chapter 5.2.2 below.

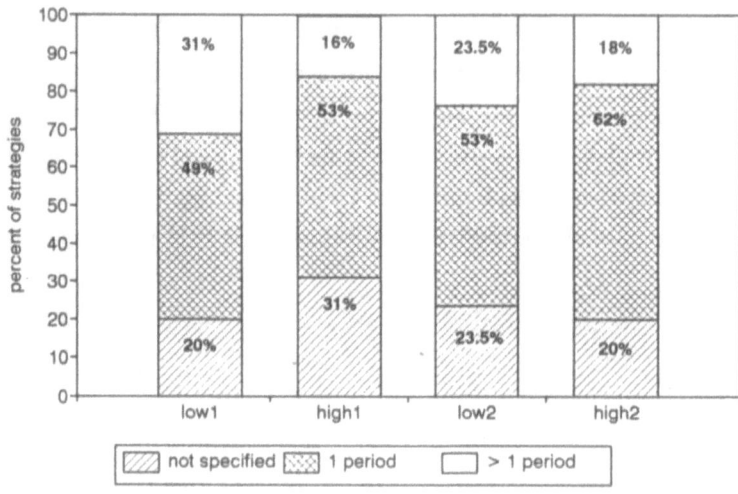

Figure 5.3: Length of the initial phases

A special rule for the last period might be necessary because the obviously optimal decision rule in the last period is to set the myopic monopoly price whatever the past history. As we can see in Figure 5.4 about 80% of the strategies do set their actual myopic monopoly prices in the last period. Some of them achieve this without specifying an end phase at all. Others apply the myopic monopoly price rule just in the last period. In quite a few strategies, however, an end phase of more than one period is specified. The maximum length observed is five periods but the endphases are typically not longer than two periods. Some of the strategies with an end phase of more than one period set the actual myopic monopoly price during the whole end phase but most of them decrease their prices before they set the monopoly price in the last period. Figure 5.5 yields the percentages of all low and all high cost strategies in each tournament which do not specify an end phase at all, which specify an end phase of one period, and which specify one of more than one period. Most (although not all!) of the strategies which do not set the actual myopic monopoly price in the last period do not specify any end phase at all.

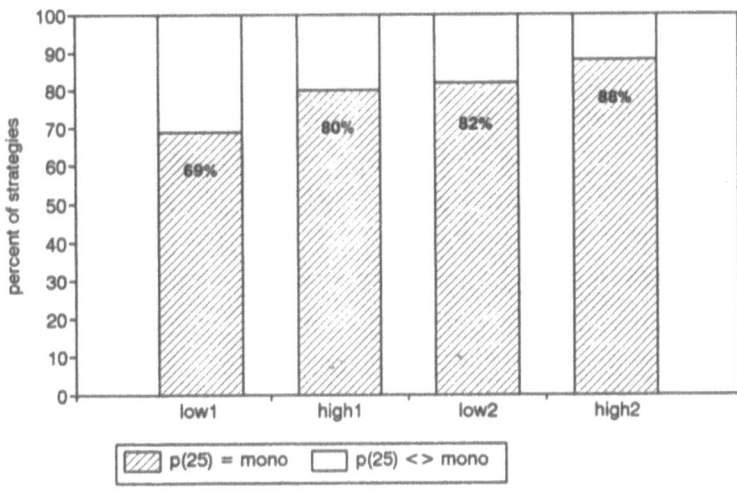

Figure 5.4: Percentage of strategies setting the myopic monopoly monopoly price in the last period

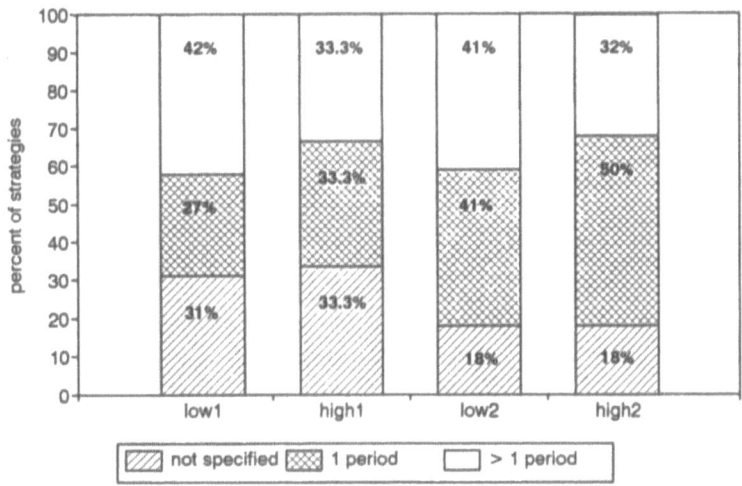

Figure 5.5: Length of the end phases

58

5.2.2 Classification of the strategies

In the following, our analysis will concentrate on the main phases of the strategies. We take the phase structure as such into account only where it seems to be of importance (e.g. a long initial phase). Because of the huge variety, we want to start out with a rough classification of the main phases which is partly inspired by an essay on bounded rationality by Selten (1990). Selten argues that according to empirical evidence a typical strategy used by a boundedly rational agent is "casuistic" in the following sense: A more or less complex system of case distinctions based on simple criteria determines which simple decision rules are employed. An alternative to a casuistic strategy is a "unified" strategy where one more or less complex formula computes the decision uniformly for all situations which may arise. The formula may be a function of past observations. A unified strategy is, according to Selten, more appealing from the point of view of full rationality.

Figure 5.6 reports the percentages of unified and casuistic strategies among the strategies submitted for each cost type and tournament. The portion of about 40% unified strategies is surprisingly high.[11] However, some of the unified strategies are not "genuinely" unified. A genuinely unified strategy is blending different kinds of considerations in one formula. If we excluded from the class of unified strategies the strategies always fixing the actual myopic monopoly price and strategies always setting a fixed price there would remain among all strategies of each cost type in each tournament only the following percentages of genuinely uniform strategies: 20% of low cost and 27% of high cost strategies in the first tournament and 29% of low cost and 32% of high cost strategies in the second tournament.

---

[11] Note that several participants told us that they know about Axelrod's (1984) tournaments for the repeated prisoner's dilemma and that the extremely simple "tit–for–tat" strategy won them. Therefore, they intended to develop a strategy with a similarly easy structure. A very short description of Axelrod's tournaments is given in Chapter 7.

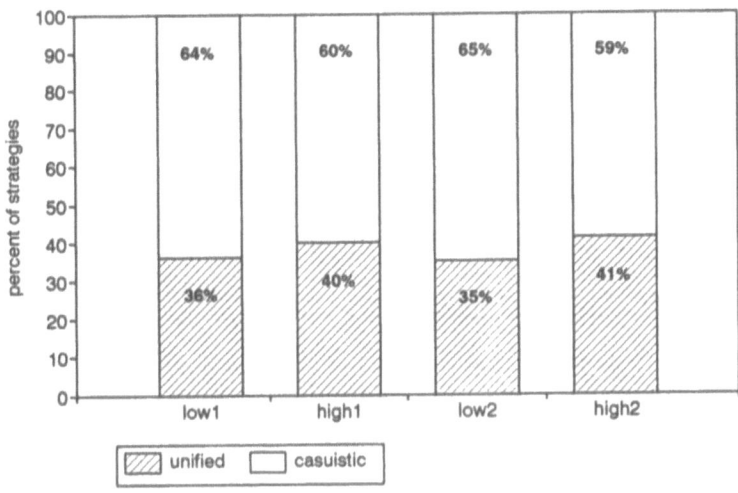

Figure 5.6:   Percentage of unified and casuistic strategies

In addition to the classification into strategies with a casuistic or unified structure, we ask which strategies involve somehow a notion of cooperation. We understand that many casuistic strategies define implicitely a "cooperative" price which is in most cases the myopic monopoly price. These strategies check in each period (of the main phase) if the opponent has behaved cooperatively in the previous period (or periods). If this is the case they also cooperate. Otherwise, they react to the deviation. The reaction is sometimes of the "measure–for–measure" – type, in the sense that a small deviation leads to a small reaction while a big deviation leads to a big reaction. Sometimes, the reaction is getting sharper with a higher degree of deviation. There are some strategies which react to any deviation by setting an extremely low fixed price. Often, the reaction as such is casuistic. We call the described strategies "cooperative" if they satisfy three additional requirements:

First, cooperation should be defined as a price not below the lower myopic monopoly price of the two firms in the market. It might be the myopic (lower) monopoly price for the actually given partition of the market demand potential or a fixed price of at least 128.50. (128.50 is the myopic monopoly price of the low cost firm if its demand potential is 200). For the high cost firm we consider also the rule "myopic monopoly price − 7" as cooperative because it specifies the myopic monopoly price of the low cost firm in the case that both firms have a demand potential of 200.

Secondly, a cooperative strategy should not start with a very aggressive initial phase; i.e. we allow the initial prices not more than twice to be below 128.50.

Thirdly, a cooperative strategy should not include any price disturbances to irritate the opponent. We would also like to require something like a moderate reaction to deviation but this cannot be clearly recognized for all possible situations from the structure of all strategies.

We also classify unified strategies as cooperative, distinguishing, however, between two possible types: Unified cooperative strategies might be with, or without reaction to deviation from cooperation. A unified cooperative strategy without reaction to deviation is, for example, one which always plays its actual myopic monopoly price, or the actual lower myopic monopoly price of the two firms in the market, or a fixed price between 128.50 and 135.50. A unified cooperative strategy reacting to deviation from cooperation sets its actual myopic monopoly price or approaches it in the case that the opponent has fixed its myopic monopoly price in the previous period. For both types of unified cooperative strategies, we also require that they do not show a very aggressive initial phase.

In contrast to the strategies with the described cooperative structure, we call all strategies where we do not recognize such a pattern of behavior "competitive" strategies. We are aware of this language use possibly being misleading; competitive should simply express that a strategy does not involve a (really cooperative) notion of cooperation. For example, a strategy always imitating the opponent's price of the previous period would not be classified by our scheme as cooperative although it may show a cooperative behavior against a cooperative opponent. Of course, the classification is to some extent arbitrary. It is to a certain degree a matter of descretion which prices, for example, are considered as cooperative.

Table 5.9 shows how the low and high cost strategies of each tournament are classified. We see that about half of all strategies exhibit a cooperative structure. This explains why we can observe so many cooperative markets in both tournaments.

Our definition of a cooperative strategy requires a not very aggressive initial phase in the sense that a price below 128.50 should not be set more than twice. Do, actually, cooperative strategies start with a price below 128.50? As we have seen that an initial phase is typically not longer than one priod, we restrict ourselves to consider only the starting price in the first period. We ask how many cooperative and competitive strategies fix in the first period a price below 128.50 and how many fix a price above or equal to 128.50. The frequencies for both tournaments are given in Table 5.10. We see that most of the cooperative strategies start with a price of at least 128.50 while most of the competitive strategies start with a price below. A χ^2 test applied to each of the 2x2–tables of Table 5.10 leads to reject, for each cost type in each tournament, the null hypothesis that equally many strategies of each class start with a price of at least 128.50 as start with a price below 128.50. The significance level is 1%.

We might also check for classification and end phase behavior. However, no systematic difference in end behavior can be recognized between cooperative and competitive strategies.

Table 5.9

Classification of the strategies in first and second tournament

		1st tournament (45 strategies)		2nd tournament (34 strategies)	
		low cost	high cost	low cost	high cost
cooperative	unified	9	12	6	8
	casuistic	14	14	9	9
		23 (51%)	**26 (58%)**	**15 (44%)**	**17 (50%)**
competitive	unified	7	6	6	6
	casuistic	15	13	13	11
		22 (49%)	**19 (42%)**	**19 (56%)**	**17 (50%)**

Table 5.10

Cooperativeness–competetiveness classification of the strategies and initial prices (p)

Low cost strategies in first tournament

initial price

		p≥128,50	p<128,50
classification	cooperative	21	2
	competitive	4	18

Low cost strategies in second tournament

initial price

		p≥128,50	p<128,50
classification	cooperative	12	3
	competitive	3	16

High cost strategies in first tournament

initial price

		p≥128,50	p<128,50
classification	cooperative	25	1
	competitive	7	12

High cost strategies in second tournament

initial price

		p≥128,50	p<128,50
classification	cooperative	16	1
	competitive	6	11

Tables 5.11 for the low cost strategies and 5.12 for the high cost strategies give us some further information about the strategies classified as cooperative. Most of them define cooperation in terms of the myopic monopoly price. As the myopic monopoly price is determined as a function of the actual demand potential which is dependent on all previous prices we might consider it as a dynamic definition of cooperation. How is deviation from cooperation measured? Some strategies check if their actual demand potential is satisfying a certain aspiration level. This aspiration level corresponds, typically, to the partition of demand potential resulting in the case that each firm sets in each period its myopic monopoly price, i.e. a demand potential of 207 for the low cost firm and of 193 for the high cost firm. Some strategies are tolerant for small deviations. There are also strategies which judge the opponent's cooperativeness by considering its price chosen in the previous period. This price is compared to the opponent's myopic monopoly price, or to one's own price in the previous period, or to a fixed price. Some strategies take both demand potential and past prices into account. The fact that many strategies require a certain demand potential points to a more static view of the game. To a large extent, it makes the dynamic character of the myopic monopoly price definition of cooperation ineffective. Note also that most of the strategies only consider the prices of the previous period. Exceptions are only two low cost strategies and one high cost strategy, all participating in both tournaments: They take all past prices (opponent's and own) into account to measure the opponent's "degree of cooperativeness" or "reputation" (participants' diction).

Table 5.11

Low cost strategies in first and second tournament

COOPERATIVE STRATEGIES

structure	reaction to deviation	notion of coop.	deviation from cooperation measured by			strategy no. in first tournament	strategy no. in second tournament
unified	no	mono				14, 22, 28	16, 28
		$p^0=130$				44	
		$p^0=132$				7	
	yes	mono	opponent's price			13, 23, 30, 36	13, 18, 23, 30
casuistic	yes	mono	demand potential	$D^0=195$		6, 34	34
				$D^0=205$			6, 14
				$D^0=207$		8, 10, 39	8
			price of opponent compared to	mono		12, 19, 31	12, 31
				fixed p		1	
				own p		26, 38	3, 26
			demand potential 207 and opponent's price compared to own price			35	
	fixed price	demand potential	$D^0=200$ $p^0=128.5$			42	
			$D^0=175$ $p^0=132$				7
			$D^0=207$ $p^0=132$			18	

Note : mono = actual myopic monopoly price
D^0 = fixed level of demand potential considered as cooperative
p = price
p^0 = fixed level of p considered as cooperative
coop. = cooperation

Table 5.12

High cost strategies in first and second tournament

COOPERATIVE STRATEGIES

structure	reaction to deviation	notion of coop.	deviation from cooperation measured by		strategy no. in first tournament	strategy no. in second tournament
unified	no	mono			3, 16, 22, 26, 28, 38	16, 26, 28
		lower mono			7, 14	7
	yes	mono	opponent's price		13, 23, 30, 36	13, 18, 23, 30
casuistic	yes	mono	demand potential	$D^0 = 190$	39	
				$D^0 = 191$		6
				$D^0 = 193$	8, 9, 10, 35	8, 14
			price of opponent compared to	mono	12, 19, 31, 33	3, 12, 31, 33
				fixed p	1	
		mono–7	demand potential	$D^0 = 195$	34	34
		fixed price	demand potential	$D^0 = 193$ $p^0 = 128.5$	42	
				$D^0 = 175$ $p^0 = 130$	15	15
				$D^0 = 93$ $p^0 = 132$	18	

Note : mono = actual myopic monopoly price
        D^0 = fixed level of demand potential considered as cooperative
        p   = price
        p^0 = fixed level of p considered as cooperative
        coop. = cooperation

Tables 5.13 and 5.14 characterize each strategy classified as competitive by its most important feature. There is a lot of variation among these strategies. However, there are some strategies with a common characteristic: They are "past price oriented". Many of them make their decisions dependent only on the prices observed in the previous periods. Some additionally take other variables into account. Among the strategies which are not past price oriented are such which play a fixed or increasing fraction of their actual myopic monopoly price. Others are open loop strategies which set in each play a fixed sequence of prices or always a fixed noncooperative price (i.e. a price of 71). There are some strategies which would be classified as cooperative if they did not violate at least one of the requirements. One participating team has submitted for both cost types and both tournaments the subgame perfect equilibrium rule (and is fairly successful in the first tournament). Another team is participating in both tournaments with strategies requiring a certain profit per period. Their strategies use random prices to initiate a search process for a price satisfying the aspiration level for the short–run profit. If after some iterations no price is found the aspiration level is reduced. Another participant is using random numbers in his price decision, but only small ones just to introduce trembling: His price decision is always to set the opponent's price of the previous period disturbed by a small random influence which is always positive.[12] A pair of rather simple unified strategies does very well in both tournaments: For each cost side, the decision rule is a smooth formula computing a weighted average of the actual myopic monopoly price, the production costs and the opponent's past price.

Let us resume that about half of the strategies have a rather uniform structure. We classify them as cooperative strategies because they involve a notion of cooperation. Among the other strategies, which we denote as competitive, there is another relatively homogeneous group: past price oriented strategies. A past price oriented strategy may show a cooperative behavior against a cooperative opponent. But it is only imitating. It is not actively cooperative. The behavioral rules of the other competitive strategies vary considerably.

---

[12] How these two strategies with random numbers arc played in the tournament is described in the introduction to Chapter 5.1 above.

Table 5.13

Low cost strategies in first and second tournament

COMPETITIVE STRATEGIES

structure	price decision		strategy no. 1st round	strategy no. 2nd round
unified	always 71		3, 16	
	subgame perfect equilibrium price for actual subgame		25	25
	75% of mono		17	
	80% of mono			17
	average of prices in previous period		40	19
	opponent's price + random (0,7)		2	2
	average of mono, prod.costs and opponent's past price		29	29
	fraction of mono increasing with time			22
casuistic	past price orientation		11,20,37,32	11,20,32
	past price orientation with attention to D		9,24,45	9, 24
	past price orientation with attention to D and profit			1
	past price orientation with attention to D, own type of reaction in previous period and "ideal price" increasing with time			5
	"ideal price" increasing with time, reaction to deviation		5,41	
	aspiration level for profit (random prices)		4	4
	checks cooperativeness of opponent but is always aggressive		33	33
	strategy with coop. structure but...	disturbances	27	27
		long non–coop. initial phase	15	15
		demanding too high a demand potential	43	
	adaptation to expected price of opponent such that partition of demand potential of 193/207 is approached			10
	fixed sequence of prices		21	21

Note: mono = actual myopic monopoly price,   coop. = cooperative,   D = demand potential

Table 5.14

High cost strategies in first and second tournament

COMPETITIVE STRATEGIES

structure	price decision	strategy no. 1st round	strategy no. 2nd round
unified	subgame perfect equilibrium price for actual subgame	25	25
	average of prices in previous period	17,40	17,19
	opponent's price in previous period	37	
	opponent's price + random (0,8)	2	
	opponent's price + random (0,4)		2
	average of mono, prod.costs and opponent's past price	29	29
	fraction of mono increasing with time		22
casuistic	past price orientation	11,20,32,44	11,20,32
	past price orientation with attention to D	24,45	9, 24
	past price orientation with attention to D and profit		1
	past price orientation with attention to D, own type of reaction in previous period and "ideal price" increasing with time		5
	"ideal price" increasing with time, reaction to deviation	5,41	
	aspiration level for profit (random prices)	4	4
	strategy with coop. structure but... — disturbances	6,27	27
	strategy with coop. structure but... — demanding too high a demand potential	43	
	adaptation to expected price of opponent such that partition of demand potential of 193/207 is approached		10
	fixed sequence of prices	21	21

Note: mono = actual myopic monopoly price
coop. = cooperative
D = demand potential

5.2.3   Some general remarks on the strategies

I had asked all participants to comment on their strategies. However, only few participants sent me more information than just a verbal description of their flow charts. To start with, let me quote here from one of the comments because it describes the typical structure of a cooperative strategy.

> "(...) At the outset the strategy is nice in the sense of practising monopoly pricing. (...) In order to avoid exploitation, there is a need to retaliate. (...) the strategy fights if the opponent was more aggressive than monopoly pricing in the previous period. Furthermore, the strategy (...) defends a potential of 193, which is the potential reached in the limit if both parties practice monopoly pricing. (...)." (participant no. 31 about his high cost strategy)

How this is actually effectuated, can be seen in the flow chart of this strategy which is shown in Appendix C.

I was told by quite a few participants that they had run simulations when they had been developing their strategies. Unfortunately, we do not know about every participant if he has used such a method or not. Thus, we cannot test whether the strategies submitted by participants having done simulations are more successful or whether they result in a typical structure. We may only say that among those who informed me that their strategies had been developed with the help of own simulations not all are very successful in the tournaments. Simulations may be very helpful but it is, for example, very difficult to create a "representative" population of opponent strategies. Furthermore, if a given strategy is compared only with similar strategies with sightly different parameters one risks very much to restrain to a very narrow range of possibilities. Some of the participants who announced to have run simulations have submitted unified strategies where parameters were "optimized" or "tested" in simulations. However, we also know about some casuistic strategies which have been developed with the help of simulations.

Typically, the structures of the low and high cost strategies submitted by a participant are very similar. In the first tournament there are seven of 45 strategy pairs which are classified differently (cooperativeness–competitiveness) for the two cost situations. Five of them are competitive on the low cost side but cooperative on the high cost side. The opposite is true for two strategies. In the second tournament, we find just one among 34 strategy pairs where the low cost strategy has a competitive structure and the high cost strategy a cooperative one. Among the strategy pairs which play cooperatively in both cost situations there are some which are casuistic in the low cost situation but unified without reaction to deviation (i.e. playing always the myopic monopoly price or the lower of both firms' myopic monopoly prices) in the high cost situation. This suggests that some participants feel as high cost firms in a disadvantageous position and behave passively rather than assertively. Participant no. 33, for example, wrote me the following about the development of the strategies which he submitted for participation in the first tournament :[13]

"(...) we had a lot of fun in experimenting with different strategies. All of us constructed some strategies and we run a small tournament. (...) As we experienced, the high cost player seemed to be somewhat more vulnerable to competitive "underpricing" than the low cost player. Therefore we wanted our strategy of the high cost player to be a kind of tit for tat in underpricing while we wanted our low cost player (initially) to be somewhat more daring in exploiting the high cost opponent."

The strategies' complexity of recall is typically one period. If it is longer, from two periods to all past periods, it applies generally to the past prices in the market. There are only two exceptions in the first tournament, checking for the demand potential in the last two periods. Figure 5.7 represents the percentages of all low cost and all high cost strategies in each tournament with memory of at most one period. Another interesting observation, going together with the bounded recall, is that no predictions about the opponent's behavior are made. Formation of price expectations, actually influencing the price decision, occurs only in one strategy pair submitted for the second tournament. These observations are of some significance because they are in contrast to many oligopoly theories which suppose that a player optimizes against his predictions of the opponent's behavior.

---

[13] Note that his low cost strategy is classified as competitive while his high cost strategy is cooperative.

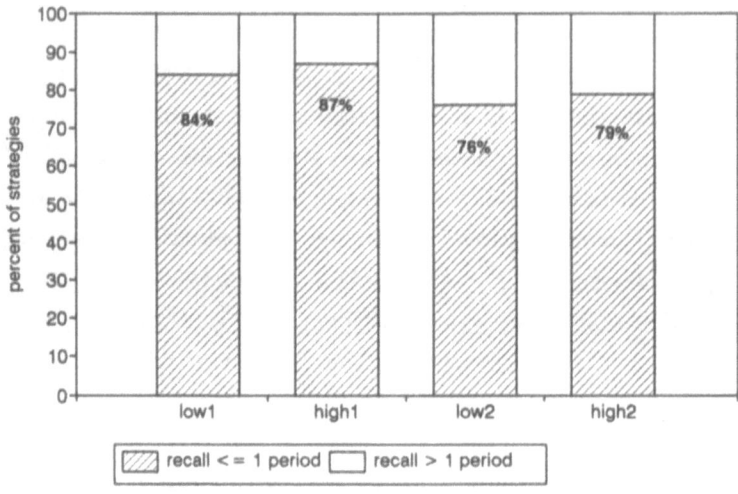

Figure 5.7:   Percentage of strategies with recall of at most
            one period

Decision rules are generally made dependent on past prices and/or the demand
potential lying above a certain aspiration level or not. An aspiration level for
profits only occurs in two participants' competitive strategies. A possible
explanation might be that realized profits result from a given combination of
demand potential and price but the same profit could also be the result of another
demand potential and price. Thus, demand potential and price variables as such
give more basic information than the short−run profit variable.

Disturbances of prices to irritate the opponent occur rarely. One participant's strategies always add small random numbers to the price decisions and two other participants' strategies disturb their prices either by playing a fixed price or by introducing an upper price limit in certain periods. The absence of such abrupt price movements in most of the strategies is astonishing because it is in contrast to what we have seen in many plays of the game–playing experiments.

It is obviously not optimal to set prices above one's actual myopic monopoly prices. This does not only lead to lower profits than playing the actual monopoly price which maximizes the short–run profit, it is also highly unfavorable in view of the future demand potential. Concavity of the short–run profit function implies that the same profit could be reached with a price below the monopoly price. Thus, a restriction for the price to be at most equal to the actual myopic monopoly price added to the pricing rules would prevent unnecessary losses of demand potential. However, about 40% of the strategies are observed to set occasionally prices above the actual myopic monopoly price in the tournaments. The exact percentages for each strategy type in each tournament are given in Figure 5.8. Our hypothesis is that these strategies do worse in the tournament than the strategies never setting prices above the actual myopic monopoly price. With a Mann–Whitney–U test, we can show that this is true on the 5% significance level for both cost types in the first tournament and the low cost strategies in the second tournament. For the high cost strategies in the second tournament, we cannot reject the null hypothesis of equal success of the strategies on a significance level of 10%.

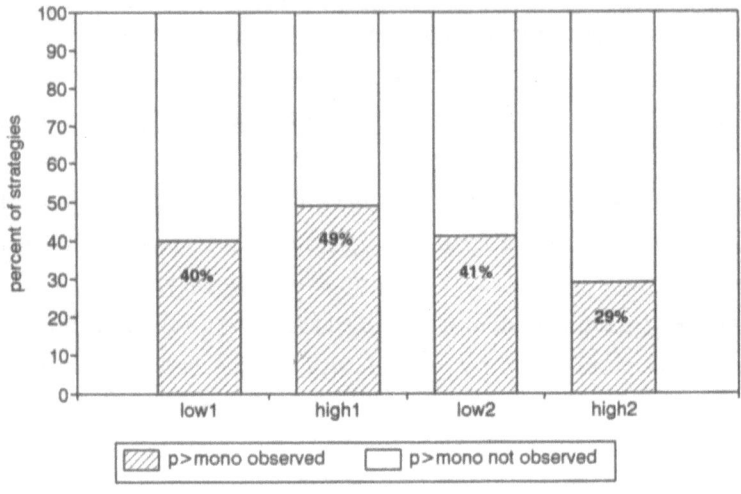

Figure 5.8 : Percentage of strategies which occasionally set prices above their actual myopic monopoly prices in the tournament

An overview of the most important characteristics of each strategy in the two tournaments is given in Tables 5.15 to 5.18. The strategies for each cost type are ordered according to their success in the corresponding tournament. We consider the tournament variation where each strategy also plays against its own counterpart.

Table 5.15
LOW COST strategies in the first tournament

rank	no.	class	unified	p(1)	initial phase	p(25) = mono	end phase	p>mono observed	periods of recall >1
1	29	comp	yes	128.50	1	yes	1	no	
2	9	comp	no	100.00	1	yes	1	no	
3	33	comp	no	114.20	1	yes	2	no	
4	43	comp	no	128.50	1	.yes	2	no	all past
5	17	comp	yes	96.38	-	yes	2	yes	
6	11	comp	no	90.00	1	no	-	yes	
7	25	comp	yes	85.07	-	yes	-	no	
8	41	comp	no	96.40	1	no	-	no	
9	4	comp	no	95-105 *	6	no	1	yes	
10	13	coop	yes	128.50	-	yes	1	no	all past
11	10	coop	no	121.50	1	yes	2	no	
12	31	coop	no	128.50	1	yes	2	no	
13	20	comp	no	125.00	1	yes	3	yes	
14	23	coop	yes	128.50	1	yes	2	yes	
15	5	comp	no	110.00	1	yes	1	no	
16	42	coop	no	128.50	2	yes	2	no	t-2
17	18	coop	no	128.50	3	yes	2	no	
18	40	comp	yes	125.00	1	no	-	no	
19	8	coop	no	128.50	1	yes	3	no	
20	34	coop	no	128.50	-	yes	4	no	
21	26	coop	no	128.50	3	no	-	no	all past
22	39	coop	no	128.50	1	yes	5	no	
23	24	comp	no	105.00	4	yes	3	yes	t-10
24	6	coop	no	130.00	2	no	3	yes	
25	35	coop	no	128.50	2	yes	2	yes	
26	30	coop	yes	128.50	2	yes	2	no	
27	27	comp	no	115.00	1	yes	2	no	t-2
28	14	coop	yes	120.00	2	yes	2	yes	
29	1	coop	no	128.50	1	no	-	no	
30	28	coop	yes	128.50	-	yes	-	no	
31	22	coop	yes	128.50	-	yes	-	no	
32	32	comp	no	132.00	3	yes	2	yes	t-3
33	15	comp	no	57.00	7	yes	1	yes	
34	38	coop	no	128.50	2	no	-	no	
35	19	coop	no	128.50	1	yes	1	no	
36	37	comp	no	125.00	2	no	-	yes	
37	44	coop	yes	128.50	4	no	-	yes	
38	2	comp	yes	77.00	1	yes	1	yes	
39	21	comp	no	80.00	1	no	1	yes	
40	36	coop	yes	130.00	1	yes	1	yes	
41	7	coop	yes	132.00	-	yes	1	yes	
42	12	coop	no	128.50	1	yes	1	no	
43	16	comp	yes	71.00	-	no	-	no	
44	3	comp	yes	71.00	-	no	-	no	
45	45	comp	no	190.00	1	no	-	yes	

* : random number
p = price, (p1) = price in first perid, p(25) = price in last period
comp = competitive, coop = cooperative
mono = actual myopic monopoly price
class = classification

Table 5.16

HIGH COST strategies in the first tournament

rank	no.	class	unified	p(1)	initial phase	p(25) = mono	end phase	p>mono observed	periods of recall >1
1	42	coop	no	128.50	·	yes	2	no	
2	29	comp	yes	135.50	1	yes	1	yes	
3	43	comp	no	135.50	1	yes	2	no	all past
4	25	comp	yes	103.76	·	yes	·	no	
5	10	coop	no	135.50	·	yes	2	no	
6	17	comp	yes	132.50	1	yes	1	yes	
7	31	coop	no	135.50	·	yes	2	no	
8	8	coop	no	135.50	1	yes	3	no	
9	39	coop	no	135.50	·	yes	6	no	
10	33	coop	no	135.50	1	yes	1	no	
11	5	comp	no	117.00	1	yes	1	no	
12	44	comp	no	130.00	1	no	·	yes	
13	23	coop	yes	135.50	1	yes	2	yes	
14	15	coop	no	130.00	1	yes	1	yes	
15	34	coop	no	128.50	·	yes	4	no	
16	7	coop	yes	128.50	·	yes	1	no	
17	14	coop	yes	128.50	·	yes	1	no	
18	18	coop	no	135.50	3	yes	2	no	
19	11	comp	no	105.00	1	no	·	yes	
20	13	coop	yes	135.50	·	yes	1	no	all past
21	30	coop	yes	135.50	2	yes	5	no	
22	19	coop	no	132.00	1	yes	1	yes	
23	27	comp	no	120.00	1	yes	2	no	t-2
24	38	coop	yes	128.50	1	yes	·	no	
25	37	comp	yes	125.00	1	no	·	yes	
26	26	coop	yes	135.50	·	yes	·	no	
27	3	coop	yes	135.50	·	yes	·	no	
28	28	coop	yes	135.50	·	yes	·	no	
29	16	coop	yes	135.50	·	yes	·	no	
30	1	coop	no	135.50	1	no	·	yes	
31	22	coop	yes	135.50	·	yes	·	no	
32	20	comp	no	125.00	1	yes	3	yes	
33	9	coop	no	120.00	1	yes	1	yes	
34	40	comp	yes	135.00	1	no	·	yes	
35	4	comp	no	95-105 *	5	yes	1	yes	
36	41	comp	no	115.20	1	no	·	yes	
37	32	comp	no	132.00	3	yes	2	yes	t-3
38	24	comp	no	99.00	4	yes	3	yes	t-10
39	35	coop	no	132.00	1	yes	2	yes	
40	36	coop	yes	130.00	1	yes	1	yes	
41	6	comp	no	120.00	2	no	·	yes	
42	2	comp	yes	88.00	1	yes	1	yes	
43	21	comp	no	80.00	1	no	1	yes	
44	12	coop	no	135.50	1	yes	1	no	
45	45	comp	no	190.00	2	no	·	yes	t-2

* : random number

p = price, p(1) = price in first period, p(25) = price in last period

comp = competitive, coop = cooperative

mono = actual myopic monopoly price

class = classification

Table 5.17
LOW COST strategies in the second tournament

rank	no.	class	unified	p(1)	initial phase	p(25) = mono	end phase	p>mono observed	periods of recall >1
1	10	comp	no	120.00	1	yes	2	no	all past
2	9	comp	no	110.00	1	yes	2	no	
3	18	coop	yes	115.65	1	yes	1	no	
4	17	comp	yes	102.80	-	yes	1	no	
5	29	comp	yes	128.50	1	yes	1	no	
6	22	comp	yes	95.52	-	yes	-	no	
7	11	comp	no	90.00	1	no	-	yes	
8	13	coop	yes	123.73	-	yes	1	no	all past
9	31	coop	no	128.50	1	yes	2	no	
10	23	coop	yes	128.50	1	yes	1	yes	
11	8	coop	no	128.50	1	yes	2	no	
12	5	comp	no	118.00	1	yes	1	no	t-2
13	20	comp	no	125.00	1	yes	3	yes	
14	24	comp	no	105.00	4	yes	3	yes	t-5
15	30	coop	yes	128.50	2	yes	2	no	
16	26	coop	no	128.50	3	no	-	no	all past
17	14	coop	no	120.00	2	no	2	yes	
18	34	coop	no	128.50	-	yes	4	no	
19	6	coop	no	132.00	3	no	3	yes	
20	3	coop	no	128.50	1	yes	1	no	
21	27	comp	no	115.00	1	yes	2	no	t-2
22	32	comp	no	132.00	3	yes	2	yes	t-3
23	1	comp	no	84.81	1	yes	2	yes	t-5
24	25	comp	yes	85.07	-	yes	-	no	
25	28	coop	yes	128.50	-	yes	-	no	
26	16	coop	yes	128.50	-	yes	-	no	
27	33	comp	no	114.20	1	yes	2	no	
28	4	comp	no	95-105 *	6	no	1	yes	
29	12	coop	no	128.50	1	yes	1	no	
30	19	comp	yes	128.50	1	yes	1	yes	
31	2	comp	yes	77.00	1	yes	1	yes	
32	7	coop	no	132.00	-	yes	1	yes	
33	15	comp	no	70.00	7	yes	1	yes	
34	21	comp	no	64.00	1	no	1	yes	

* : random number
p = price, p(1) = price in first period, p(25) = price in last period
comp = competitive, coop = cooperative
mono = actual myopic monopoly price
class = classification

Table 5.18

HIGH COST strategies in the second tournament

rank	no.	type	unified	p(1)	initial phase	p(25) = mono	end phase	p>mono observed	periods of recall >1
1	29	comp	yes	135.50	1	yes	1	no	all past
2	10	comp	no	125.00	1	yes	2	no	
3	9	comp	no	125.00	1	yes	2	no	
4	22	comp	yes	119.38	-	yes	-	no	
5	17	comp	yes	132.50	1	yes	1	yes	
6	11	comp	no	105.00	1	no	-	yes	
7	18	coop	yes	121.95	1	yes	1	no	
8	5	comp	no	130.00	1	yes	1	yes	t-2
9	8	coop	no	135.50	1	yes	2	no	
10	31	coop	no	135.50	1	yes	2	no	
11	15	coop	no	130.00	1	yes	1	yes	
12	23	coop	yes	135.50	1	yes	1	yes	
13	3	coop	no	135.50	1	no	1	no	
14	7	coop	yes	128.50	-	yes	1	no	
15	30	coop	yes	135.50	2	yes	5	no	
16	13	coop	yes	129.01	1	yes	1	no	all past
17	25	comp	yes	103.76	-	yes	-	no	
18	6	coop	no	132.00	3	no	3	yes	
19	14	coop	no	128.50	1	yes	1	no	
20	33	coop	no	135.50	1	yes	1	no	
21	27	comp	no	120.00	1	yes	2	no	t-2
22	20	comp	no	132.00	1	yes	3	yes	
23	34	coop	no	128.50	-	yes	4	no	
24	28	coop	yes	135.50	-	yes	-	no	
25	26	coop	yes	135.50	-	yes	-	no	
26	16	coop	yes	135.50	-	yes	-	no	
27	4	comp	no	95-105 *	5	yes	1	no	
28	1	comp	no	104.34	2	yes	1	no	t-2
29	32	comp	no	132.00	3	yes	2	no	t-3
30	24	comp	no	99.00	4	yes	3	no	t-5
31	19	comp	yes	135.50	1	yes	1	yes	
32	2	comp	yes	77.00	1	yes	1	yes	
33	12	coop	no	135.50	1	yes	1	no	
34	21	comp	no	71.00	1	no	1	yes	

* : random number
p = price, p(1) = price in first period, p(25) = price in last period
comp = competitive, coop = cooperative
mono = actual myopic monopoly price
class = classification

Let me mention that one participating team published its ideas about IDEAS and a description of its strategies which participated very successfully in the first tournament.[14] In one section of this publication, the authors discuss the advantages and problems of cooperation: Tacit collusion allows for both duopolists higher profits than in the subgame perfect equilibrium solution of the game. But they mention two difficulties: It is not really clear what exactly cooperation means in this game. Another difficulty lies in the choice of the degree of cooperation as there is an advantage in being a little bit less cooperative than the most cooperative. This last consideration has apparently been quite influential on the strategies which the authors submitted for participation in the first tournament. Their strategies would be classified as cooperative if they did not claim, on both cost sides, a demand potential very much in their own favor. These high claims turned out as "best" in "homemade simulations" which the authors also describe in their publication. Let me reprint the description of their strategies in their own words:

"Structure de la stratégie adoptée.

On considère qu'il existe un niveau de collaboration de référence; il correspond à la situation où chaque firme pratique son prix de monopole sur sa part de marché et que leur prix sont égaux (les parts de marchés sont donc stables). Ceci conduit à:

– un prix $p_M = 132$.
– une part de marché $D_M = 193$ pour la firme ayant des coùts élevés.
– une part de marché $D'_M = 207$ pour l'autre firme.

La stratégie consiste à obtenir une part de marché D_S, plus grande que D_M, et y pratiquer le prix de monopole $p_S = 1/2\,(D+c)$.

Pour y parvenir, on adopte la stratégie suivante (stratégie de type S). A la période t, les parts de marchés et les prix de marchés des périodes précédentes sont connues, ainsi que les parts de marchés respectives, D_t, de la période courante.

---

[14] J. Bourdieu and M Servain, DELTA au tournoi I.D.E.A.S., in: La Lettre Δ, DELTA, April 1991.

– Si $D_t \geq D_S$, alors on joue $p_t = \frac{1}{2}(D_t+c)$.

– Si $D_t < D_S$, alors, pour gagner des parts de marché, on "casse" les prix en pratiquant

$\cdot\ p_t = p_{t-1}$, quand $p_{t-1} \leq p^*_{t-1}$

$\cdot\ p_t = p_{t-1} - \beta_t(p_{t-1} - p^*_{t-1})$, quand $p_{t-1} \geq p^*_{t-1}$

avec $\beta_t = \beta_{t-1} + \alpha(p_{t-1} - p^*_{t-1})$ et $\beta_0 = 0$

où p^*_{t-1} est le prix pratiqué par l'autre joueur.

Cette stratégie permet de procéder à une "riposte graduée", puisque, dans un cas, ayant eu à la période précédente un prix inférieur au prix de l'adversaire, on se contente de reprendre le même prix, et, dans le cas contraire, on réduit le prix d'autant plus fortement que (β_t), notre prix, était élevé par rapport à celui de l'adversaire. De plus, les β_t augmentent de manière irréversible, lorsque nos prix sont eux–mêmes cassés par la firme adverse."

The parameters α, β and D_S have been selected with the help of simulations where it turned out that is is profitable to be a little more aggressive than the "average aggressivity of the environment". An argument in favor of being (moderately) aggressive can also be found in the reflections on the game of participating team no. 10. This team participated in the first tournament with strategies classified as cooperative. For the second tournament they submitted, together with the comment from which I am quoting, more aggressive strategies which we classify as competitive. They recognize that the higher a firm's demand potential, the lower a price it needs, to gain the same short–run profit:

"Die erfolgreichsten Strategien von Runde 1 waren nicht kooperativ sondern haben dauerhaft niedrige Preise gespielt. (Dies ist der Eindruck nach dem Verhalten gegenüber unserer Strategie.) Der Vorteil einer solchen Strategie ist, daß sie sich nicht anzupassen braucht: Gegenüber einem anderen "Aggressor" ist sie allemal angebracht, und ein Spieler, der systematisch nur seinen Monopolpreis (oder nur wenige Punkte darunter) verlangt, wird zum eigenen Vorteil "ausgeraubt": Der Gewinn von 75[2], den etwa Low bei einem Demandpotential von d=207 bei einem Preis von p=132 erhält, erhält Low auch für d>207 bei einem Preis p, der um so <u>niedriger</u> gewählt werden kann, je größer d ist, und zwar besonders stark am Anfang: Für d=208 etwa liefert p=123.83 den gleichen Gewinn, für d=210: p=118.43, für d=220: p=106.61 und für d=227: p=102. (...) Bei einem hohen Potential kann man sich also einen niedrigen Preis leisten, ohne gegenüber dem kooperativen Fall mit Preis 132 etwas zu verlieren."

This team's strategies for the second tournament, although classified as competitive, involve a notion of cooperation which aims at the partition of demand potential which results in the myopic monopoly solution. However, being cautious not to lose demand potential they start with a relatively low price and never fix a price above 125 in the main phase. They are very successful in the second tournament.

5.2.4 Strategy modifications after the first tournament

Eleven participants of the first tournament do not participate in the second tournament. Table 5.19 gives an overview of their structure and success in the first tournament. We cannot find out, as one might suspect, that the least successful strategies or strategies with a certain structure dropped out. Thus, we suppose that the participants had personal reasons (e.g. time shortage, no more interest) for not participating in the second round. After the first tournament I had asked all participants to return an answer sheet where they should have signed for "I (we) want to participate in the second tournament" and either "with a new pair of strategies (...)" or "with my (our) strategies of the first tournament". Participants who did not answer to this were excluded from the second tournament. New participants were not permitted.

Table 5.19

Cooperativeness–competitiveness classification and success of the strategies of the 11 participants not paticipating in the second tournament

classification and success in first tournament		# low cost strategies	# high cost strategies
classification	cooperative	6	5
	competitive	5	6
success	profit > median	5	4
	profit < median	6	7

Ten of the 34 participants who participate in the second tournament round keep their strategy <u>pairs</u> unchanged. Let us start with analyzing if the participants who submitted modified strategies had changed the structure of their strategies in a way that leads to a different cooperativeness–competitiveness classification. Has the distribution of cooperative and competitive strategies changed from the first to the second tournament? We consider only the strategies of the 34 participants who participate in both tournaments. Tables 5.20 and 5.21 show, for the low and the high cost side, how many cooperative and how many competitive strategies of the first tournament participate in the second round as strategies of the same cooperativeness–competitiveness class and how many are classified differently. We see that not many strategies change their cooperativeness–competitiveness structure in the second tournament and that among those which do, there are more which turn from cooperative to competitive than the other way round. However, a Mc Nemar χ^2 test applied to both 2x2–tables does not yield support for a significant structure change if we require a significance level of 10%.[15]

---

[15] Note that this is principally due to the fact that we have got too few observations of strategies changing their structures.

Table 5.20

Cooperativeness–competitiveness classification of low cost strategies
participating in both tournaments

2nd tournament

	cooperative	competitive
cooperative	13*	4
competitive	2	15**

1st tournament

*   5 unchanged strategies
** 6 unchanged strategies

Table 5.21

Cooperativeness–competitiveness classification of high cost strategies
participating in both tournaments

2nd tournament

	cooperative	competitive
cooperative	16*	5
competitive	1	12**

1st tournament

*   8 unchanged strategies
** 4 unchanged strategies

Our next question is: Do the strategy modifications after the first tournament lead to more or to less aggressive behavior? Therefore, we would like to make the aggressiveness of a strategy measurable and use this measure to compare the strategies of both tournaments. In the previous chapter, we have considered the average opponents' profit in the plays against a strategy as a measure of its aggressiveness. Unfortunately, this measure is highly dependent on the population of opponent strategies. Thus, it would not be adequate a measure to compare the aggressiveness of each participant's strategies in first and second tournament. We might, however, consider the profit of a strategy against a reference strategy, or the profit of a reference strategy against this strategy as an indicator for the aggressiveness of the strategy and use it as an instrument to check for the effect of the strategy modification.

Let us consider an artificial strategy which does not participate in the tournaments: a price imitator which imitates in each period the opponent's past price. It starts in the first period with its own myopic monopoly price. This price imitating behavior might be viewed as a straightforward translation of the tit–for–tat principle into the considered game situation.[16] We let all permanent participants' low cost and high cost strategies of each tournament play against the price imitating strategy. The average profits of the imitator realized against each population are reported in Table 5.22. We see that its profit against all low cost strategies of the second tournament is higher than the profit against all low cost strategies of the first tournament. The opposite is true for its profits against the high cost strategies. This hints to the low cost strategies having become on average less aggressive and the high cost strategies more aggressive.

---

[16] We have chosen this reference strategy because we have seen that its profit against a strategy participating in the second tournament reflects the average opponents' profits of this strategy. See Chapter 5.3 below.

Table 5.22

Average profit of the price imitating strategy against
the 34 strategies participating in both tournaments

imitator	average profit against 34 opponents of	
	first tournament	second tournament
low cost	147 470	144 372
high cost	91 278	96 278

We want to distinguish now between a general aggressiveness of a strategy which finds expression in the degree of exploitation of "unprotected" cooperative behavior and a strategy's "protection" against aggressive opponent behavior. Therefore, we base the following analysis on two measures which should capture these two aspects of a strategy's aggressiveness:

- the profit of the strategy always fixing its myopic monopoly price (myopic monopoly strategy) in the play against a considered strategy

- the profit of a strategy in the play against the strategy always playing the subgame perfect equilibrium solution for the remaining subgame.

We interprete these measures in the following way: If we observe that a participant's strategy in the second tournament allows the monopoly strategy to make a higher profit than in the first tournament, we conclude that this strategy, having become more cooperative towards an unprotected cooperative opponent, is likely to have become generally less aggressive. If a strategy's profit against the subgame perfect equilibrium strategy has increased we conclude that the strategy is likely to be better protected against an aggressive opponent. Note that we cautiously say "is likely to" because the statements need not always be true.

Table 5.23 summarizes for the low and high cost strategies the information given by these two measures. A "+" represents an increase and "−" a decrease of the considered profit and "=" means that the considered profit has not changed at all. We recognize that there is no clear direction of modification of the strategies after the first tournament, especially not for the high cost strategies. On the low cost side it is, however, striking that 20 of the 23 modified strategies show a better protection against the aggressive opponent. But we cannot recognize a clear tendency towards or away from a generally more cooperative behavior: 10 strategies become more cooperative, 9 less cooperative and 4 do not change their behavior towards the myopic monopoly strategy. On the high cost side, half of the strategies behave more aggressively while about one quarter behaves more cooperatively and another quarter does not change behavior at all against the myopic monopoly strategy. 11 of the 22 modified high cost strategies improve their protection against the subgame perfect equilibrium strategy while 9 strategies do worse and 2 do neither better nor worse against the subgame perfect equilibrium strategy.

Although the result is not very clear, we conclude that on average there is, on the low cost side, a trend towards better protection against aggressive opponents while on the high cost side there might be a slight tendency towards a generally more aggressive behavior.

Table 5.23

From first to second tournament: Profit of a strategy against the subgame perfect equilibrium strategy (SPE) and the profit of the myopic monopoly strategy (mono) against this strategy

(+ : increase, – : decrease, = : unchanged)

23 low cost strategies modified in the second tournament:

own profit against SPE

		+	–	=
	+	7	2	1
profit of mono	–	9		
	=	4		

22 high cost strategies modified in the second tournament:

own profit against SPE

		+	–	–
	+	3	2	
profit of mono	–	4	7	
	=	4		2

5.3   Success of the strategies

We would like to know what determines the success of a strategy. In Chapter 5.1.4 we have shown that both the most successful and the least successful strategies are more aggressive (measured by the opponents' profits) than the majority of the strategies. At the same time, we suspect that the success of a strategy is linked to its structure (cooperative or competitive). Inquiring Tables 5.17 and 5.18, we get the impression that the strategies with a cooperative structure are located in the middle range of the ranking lists. This implies that the dispersion of the profits of the cooperative strategies is smaller than that of the competitive strategies.

To test for the smaller dispersion of profits of the cooperative strategies, we apply the following quartile test:[17] We order the strategies of each cost type and each tournament according to their average profit over all plays (ranking lists of profit). Then, we divide them into profit "quartiles" such that quartile Q1 contains the most successful strategies and Q4 the least successful strategies. As our hypothesis is that the cooperative strategies are more in the middle than in the extremes of the ranking, i.e. in the quartiles Q2 and Q3, we combine these quartiles to the "inner" quartiles and Q1 and Q4 to the "outer" quartiles. Note that the number of participants is in no tournament dividable by four. We have done the division into quartiles such that the inner quartiles have been chosen larger than the outer quartiles.[18] We count the number of cooperative and competitive strategies in inner and outer quartiles. These frequencies are presented in the 2x2–tables of Table 5.24. We apply a χ^2 test for the null hypothesis of equal proportion of cooperative and competitive strategies in inner and outer quartiles. We may, for each cost type in each tournament, reject the null hypothesis on the 5% significance level (one–sided test) in favor of our hypothesis that strategies classified as cooperative are more likely in the inner quartiles.

---

[17] This test is similar to the quartile–median test for the success and the aggressiveness of the strategies described in Chapter 5.1.4.

[18] Instead of omitting no.23 in the first tournament with 45 participants, a procedure suggested by Lienert (1973), we add no.23 to Q2+Q3 and, again, instead of omitting no.9 and 26 in the second tournament with 34 strategies, we add them to the inner quartiles.

Table 5.24

Cooperativeness – competitiveness classification and success of the strategies

Low cost strategies in first tournament

classification

	competitive	cooperative
outer profit quartiles	15	7
inner profit quartiles	7	16

success

Low cost strategies in second tournament

classification

	competitive	cooperative
outer profit quartiles	12	4
inner profit quartiles	7	11

success

High cost strategies in first tournament

classification

	competitive	cooperative
outer profit quartiles	13	9
inner profit quartiles	6	17

success

High cost strategies in second tournament

classification

	competitive	cooperative
outer profit quartiles	14	2
inner profit quartiles	3	15

success

We have shown by now that the dispersion of profit of the competitive strategies is higher than of the cooperative strategies. In connection with the former observation that the least and the most successful strategies are more aggressive than the average, we state the following hypothesis: Strategies classified as competitive are more aggressive than strategies classified as cooperative in the sense that their opponents make lower profits. To carry out a median test we group our strategies according to two criteria: First, do they belong to the class of cooperative or competitive strategies and, secondly, is their opponents' profit above or below the median of the whole sample. The resulting frequencies are given in the 2x2–tables of Table 5.25. We apply a one–tailed χ^2 test for the null hypothesis that in both classes, cooperative and competitive, equally many observations lie above and below the median. We may reject the null hypothesis on the 5% significance level for each cost type in each tournament.

We have seen so far that in both tournaments the strategies classified as competitive are more aggressive than those classified as cooperative. They tend to be in the outer profit quartiles rather than the inner ones. We may say that a competitive structure is "riskier" than a cooperative structure in the sense that the dispersion of profits of the cooperative strategies is significantly lower.

Table 5.25

Cooperativeness – competitiveness classification and aggressiveness of the strategies

Low cost strategies in first tournament

aggressiveness

	opponents' profit $\leq$ median	opponents' profit $>$ median
comperative	18	4
cooperative	5	18

classification

Low cost strategies in second tournament

aggressiveness

	opponents' profit $\leq$ median	opponents' profit $>$ median
competitive	16	3
cooperative	1	14

classification

High cost strategies in first tournament

aggressiveness

	opponents' profit $\leq$ median	opponents' profit $>$ median
competitive	16	3
cooperative	7	19

classification

High cost strategies in second tournament

aggressiveness

	opponents' profit $\leq$ median	opponents' profit $>$ median
competitive	13	4
cooperative	4	13

classification

How does this come about? Let us make a deeper analysis of the behavior of the strategies participating in the second tournament. Here we concentrate on the second tournament because the participants were then more experienced than in the first round. We consider now the profits of each strategy against three reference strategies:

- the strategy always playing its actual myopic monopoly price (the myopic monopoly strategy)

- the strategy always playing its subgame perfect equilibrium price for the remaining subgame (the subgame perfect equilibrium strategy)

- the strategy starting with its actual myopic monopoly price in the first period and then always imitating the opponent's past price (a price imitating strategy).

We divide our strategy samples of each cost type into cooperative and competitive strategies. Table 5.26 reports for each group the average rank of profits against each reference strategy. We apply a Mann–Whitney–U test to see whether the competitive strategies do significantly better or worse than the cooperative strategies against the myopic monopoly strategy, the subgame perfect equilibrium strategy and the price imitating strategy. The significance levels of the two–tailed test are also given in Table 5.26. We see that the competitive strategies do on the .5% significance level better against the myopic monopoly strategy than the cooperative strategies. They "exploit" this "unprotected" cooperative strategy. The competitive strategies are on average also better against the subgame perfect equilibrium strategy although this is significant only on the low cost side if we require a significance level of 10%. We might consider the competitive strategies as better protected against such a relatively aggressive strategy. We cannot find, however, a remarkable difference of the profit of the competitive and cooperative strategies in the play against the price imitating strategy.

Table 5.26

Average profit ranks of cooperative (coop) and competitive (comp) strategies in the plays
against the myopic monopoly strategy (mono), the subgame perfect equilibrium strategy (SPE),
and the mere price imitating strategy starting with its monopoly price (imitator).

profit against	LOW COST strategies		α	HIGH COST strategies		α
	average profit rank * of strategies classified as			average profit rank * of strategies classified as		
	comp	coop		comp	coop	
mono	12.6	23.7	.005	11.4	23.6˙	.005
SPE	12.7	23.5	.005	14.8	20.2	**
imitator	18.4	16.4	**	16.1	18.9	**

α : Significance level for a two–tailed Mann–Whitney–U test to reject the null hypothesis
    of equal profit of cooperative and competitive strategies against the reference strategy.
* : 1 = highest profit
** : not significant on the 10% level

These observations, together with the fact that the cooperative strategies are more likely to make profits in a middle range while the competitive strategies are more likely to make either very high or very low profits, are in line with the following observation. Table 5.27 reports rank correlations (Spearman) between success of a strategy in the tournament and the profit against the monopoly, the subgame perfect equilibrium and price imitating strategy. We see that the success against the price imitator is a rather good indicator for the success in the tournament while the profits against the other two reference strategies are less significantly correlated with the success in the tournament. The price imitating strategy represents to some extent the behavior of the whole population of opponent strategies. The success of a strategy in the play against the price imitating strategy may be considered to be a rather reliable indicator for the strategy's success in the tournament.

Table 5.27

Spearman rank correlation coefficients (r_s)* between the strategies' success in the second tournament and their profits against the monopoly (mono), the subgame perfect equilibrium (SPE), and price imitating (imitator) strategy.

	LOW COST strategies		HIGH COST strategies	
	r_s	α	r_s	α
success – profit against mono	.35	.005	.15	**
success – profit against SPE	.48	.001	.24	**
success – profit against imitator	.79	.000	.59	.001

* : computed with STATGRAPHICS version 2.6
** : not significant on the 10% level
α : significance level of r_s

Note that not only the profit of a strategy in the play against the price imitating strategy is a good indicator for the success of this strategy in the second tournament but also the profit of the price imitator itself against the considered strategy is a good indicator for the opponents' profit in the second tournament. Rank correlation of the opponents' profit of a strategy and the price imitator's profit in the play against this strategy are .8767 on the low cost side and .5896 on the high cost side. These coefficients are significant on the .5% significance level.

To conclude, let us interpret the main results of this chapter in the following way: There are two principal features which are important for the success of a strategy. The first is being cooperative or competitive. The second finds expression in the realized profit in the play against a price imitating strategy. Cooperative strategies are very likely to be in the middle range of the success rankings, while competitive strategies tend to be either at the top or the bottom. By its construction, a cooperative strategy generally cooperates with other cooperative strategies. Competitive strategies, on the contrary, typically are more aggressive. Here the second feature is of importance: If a competitive strategy is just a little aggressive it can not only exploit the unprotected cooperative strategies but also make high profits against cooperative strategies tolerating small deviations from cooperation without reacting sharply and against price imitating strategies which start "nicely" (i.e. which fix in the first period a price to be considered as cooperative). Then, a competitive strategy is very successful in the tournament. A competitive strategy risks, however, to exceed the limit and to be overly aggressive. Then, it will make relatively low profits against price imitating strategies as well as against cooperative strategies reacting to deviation from cooperation. This leads to being unsuccessful in the tournament in spite of the exploitation of the unprotected cooperative strategies.

5.4   An evolutionary tournament

In this chapter we want to see what happens if we let the strategies of the second tournament play against each other in an evolutionary tournament based on the discrete replicator dynamics. We are interested if among the strategies there are pairs which are "evolutionarily stable". In evolutionary game theory, the concept of an evolutionarily stable strategy (ESS) is defined as a "strategy such that, if all members of a population adopt this strategy, then no mutant strategy could invade the population under the influence of natural selection." (Maynard Smith (1982))

Before we describe the evolutionary process let us analyze the 34x34 profit bimatrix resulting from each low cost strategy playing against each high cost strategy. We want to consider it as the payoff matrix of a 34x34 two–person normal–form game where one player is a low cost player L and the other is a high cost player H. Each has to chose one of the 34 strategies for his cost side. Let l_i be the low cost strategy submitted by participant no.i, and h_i the high cost strategy submitted by participant no.i. Note that we distinguish between l_i and l_j, or h_i and h_j even if participants i and j have submitted, for the considered cost side, strategies with exactly the same rules. The low cost player L choses a strategy $l \in \{l_i/i = 1,...,34\}$ and the high cost player H choses a strategy $h \in \{h_i/i=1,...,34\}$. Let $g^L(l_i,h_j)$ denote the profit of a low cost firm with strategy l_i in the play against a high cost firm with strategy h_j and $g^H(l_i,h_j)$ denote the profit of a high cost firm with strategy h_j in the play against a low cost firm with strategy l_i. the payoff matrix is then described by $(g^L(l_i,h_j),g^H(l_i,h_j))_{1 \leq i,j \leq 34}$. In the following, I call this the "34x34 normal–form game".

What about pure strategy equilibria in this game? We find that there are two strategy pairs for the low and high cost situation which are best replies to themselves. One is, of course, the subgame perfect equilibrium strategy pair which was submitted by participant no.25. The other equilibrium strategy pair is also one where both strategies were submitted by the same participant, no.4, and it has the same structure on both cost sides: It searches in each period a price leading to a short–run profit not below a certain level. If after some iterations no such price is found, then the profit aspiration level is reduced. As we want to compare these two equilibria as possible candidates for evolutionarily stable strategies, we should point out that the subgame perfect equilibrium strategies show the advantage that they are not only best replies to themselves but also, on both cost sides, best replies to seven other participating strategies. To play the subgame perfect equilibrium strategy is a best reply to "unprotected" cooperative strategies as, for example, the strategy always fixing its actual myopic monopoly price. The other strategy pair which is an equilibrium in pure strategies of the 34x34 normal–form game, however, makes higher profits in the play against itself than does the subgame perfect equilibrium strategies make against themselves.

Will we find stabilization in an evolutionary process based on this game? If yes, will it stabilize one of the pure equilibria, or will we find a mixed equilibrium?

We construct a discrete replicator process based on the 34x34 normal–form game. We consider $r = 0, ..., 100\,000$ iterations. Suppose that we have a (large) population of low cost firms of which each follows some low cost strategy l and a (large) population of high cost firms of which each follows some high cost strategy h. Assume that each firm matches once, in each iteration, a firm of the opposite cost type to play the dynamic duopoly game. In round 0, each strategy of each cost type is played by the same percentage of firms of the given cost type. In the following, we shall call the percentage of the population by which a given strategy is followed as the "weight" of this stategy in the population. This weight represents the probability to meet an opponent following this strategy. Thus, the payoff of a strategy in a considered round is its expected profit given the weights of the opponent strategies. Let $w^r(l_i)$ be the percentage of low cost firms playing strategy l_i in round r, or the weight of low cost strategy l_i in round r, and let $w^r(h_i)$ be the percentage of high cost firms playing strategy h_i in round r, or the weight of high cost strategy h_i in round r. The payoff of low cost strategy l_i in round r, $\Pi^r(l_i)$, and the payoff of high cost strategy h_i in round r, $\Pi^r(h_i)$, are then given by:

$$\Pi^r(l_i) = \sum_j w^r(h_j)\, g^L(l_i, h_j)$$

$$\Pi^r(h_i) = \sum_j w^r(l_j)\, g^H(l_j, h_i)$$

(9)

Survival of a strategy in an evolutionary process depends on "fitness". In our case, fitness of a strategy is measured by its expected payoff. The evolution dynamics is modelled as a replicator dynamics which is modified by recurrent mutations. Recurrent mutations prevent dying out of strategies: Again and again, with a small rate, strategies spontaneously enter the population.[19] The replicator dynamics, modified by recurrent mutations, is modelled by the following system of equations which determine the new weight of a strategy in round r+1 as a function of the weights of all strategies in round r and then modify it according to recurrent mutations:

---

[19] See for example Roughgarden (1979).

$$\omega^{r+1}(l_i) = \frac{\Pi^r(l_i)}{\displaystyle\sum_k \sum_j w^r(l_k) \, w^r(h_j) \, g^L(l_k,h_j)} \, w^r(l_i)$$

$$\omega^{r+1}(h_i) = \frac{\Pi^r(h_i)}{\displaystyle\sum_k \sum_j w^r(h_k) \, w^r(l_j) \, g^H(l_j,h_k)} \, w^r(h_i)$$

$$(10)$$

$$w^r(l_i) = \omega^r(l_i) \, (1 - 34 \cdot 10^{-7}) + 10^{-7}$$

$$w^r(h_i) = \omega^r(h_i) \, (1 - 34 \cdot 10^{-7}) + 10^{-7}$$

$$\text{for } i = 1,...,34, \quad r = 1,...,100\,000$$

where $\omega^r(l_i)$ and $\omega^r(h_i)$ denote the weights of the strategies before they are modified by recurrent mutations. Technically, the mutations are modelled by taking away from each strategy a part of $34 \cdot 10^{-7}$ of its weight and adding linearily 10^{-7} to each strategy's weight. As the weights of the strategies of each cost type add up to 1 we take away, over all strategies of a cost type, exactly the same amount that we redistribute equally afterwards. A redistribution, or a minimum weight of 10^{-7}, may be considered as reasonable because it is very small but still within the exactness range of the computer.

The result of the evolution tournament: We find that after 100 000 rounds of the described evolution process there are only two strategies on each cost side which have "survived" with a high weight. They are not the equilibrium strategies described above. They are low cost strategies no.18 and 27 and high cost strategies no.17 and 31. Figure 5.9 for the low cost side and Figure 5.10 for the high cost side show the evolution of the weights of the two widespread strategies as well as of two other strategies on each cost side. Note that for technical reasons, the figures indicate the weights in every hundredth iteration only. To start with, let us consider on each cost side the two strategies with the high weights. We recognize in Figures 5.9 and 5.10 that their weights show cyclical movements and convey the impression of approaching limit cycles. Therefore, we will compute the time average of their weights. Table 5.28 reports for each strategy of each cost side the average weight over all 100 000 periods. On the low cost side, about 72% of the firms follow strategy no.18 while 24% follow strategy no.27 on time average. On the high cost side, we have 57% of the firms playing strategy no.17 and 38% playing strategy no.31. On neither cost side there is any other strategy played by at least 5% of the population on time average.

We might suppose that the two widespread strategies on each cost side might form a mixed equilibrium of the 34x34 normal–form game on which the evolution process is based. To analyze this, let us consider first the 2x2 game which results if we restrict the strategy sets for the low and the high cost player to the two surviving strategies on each cost side and compute the mixed strategy equilibrium of the 2x2 game. (There is no equilibrium in pure strategies of this restricted game.) Then, we have to check whether it is true that among all 34 strategies of each cost side there is no pure strategy which makes against the mixed strategy of the opposite cost side a higher profit than the equilibrium profit of the 2x2 restricted game. We find, however, that on both cost sides there are pure strategies which do better. Especially, low cost strategy no.15 makes against high cost strategies 17 and 31, played with their equilibrium probabilities, a largely higher profit than strategies no.18 and 27 do. The four widespread strategies do not form an equilibrium in mixed strategies of the 34x34 game.

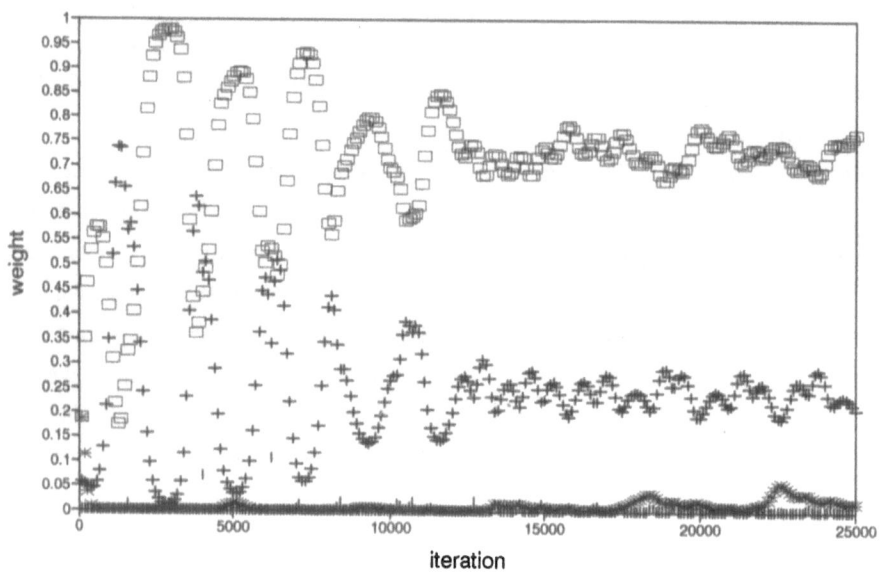

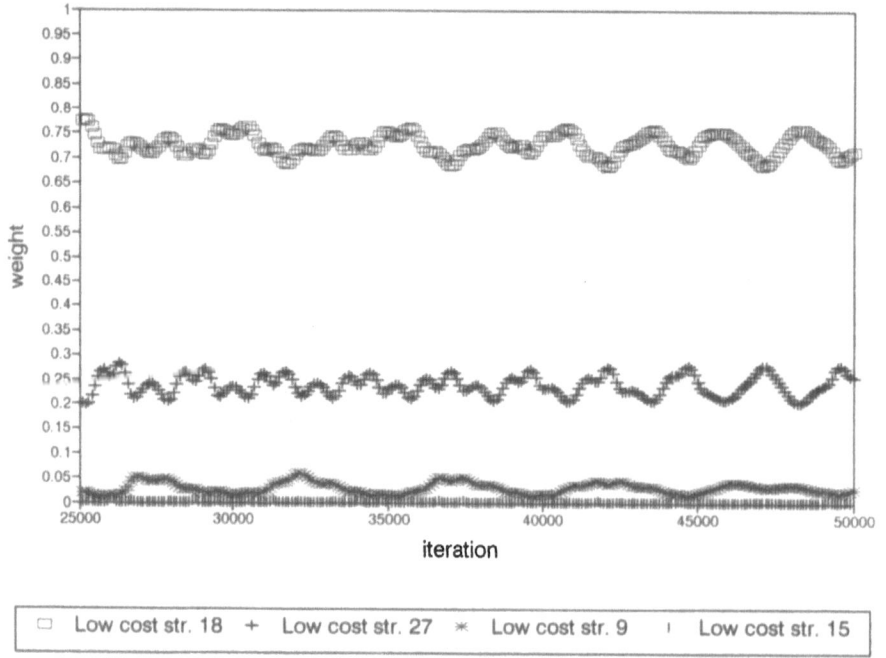

□ Low cost str. 18	+ Low cost str. 27	✳ Low cost str. 9	ı Low cost str. 15

Figure 5.9 (Part I):   Weights of selected low cost strategies in an evolutionary process with 100 000 iterations

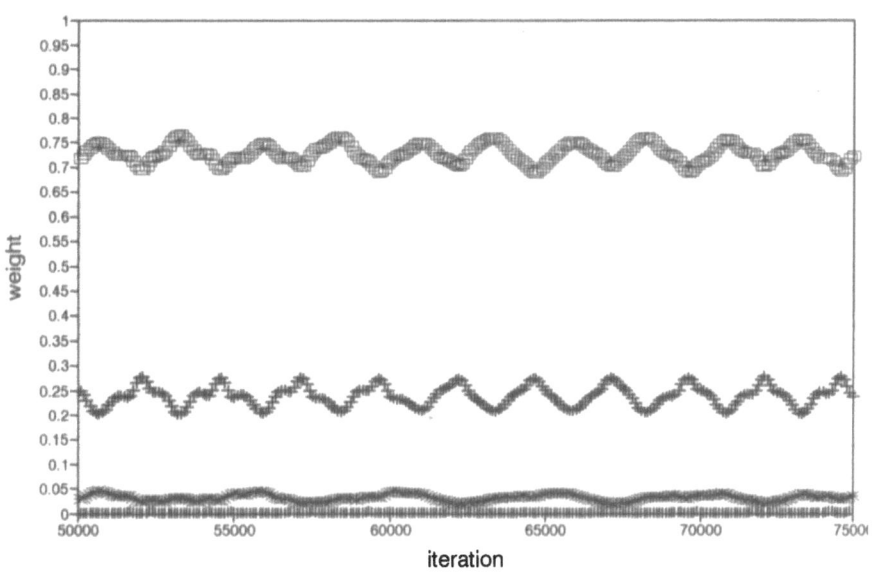

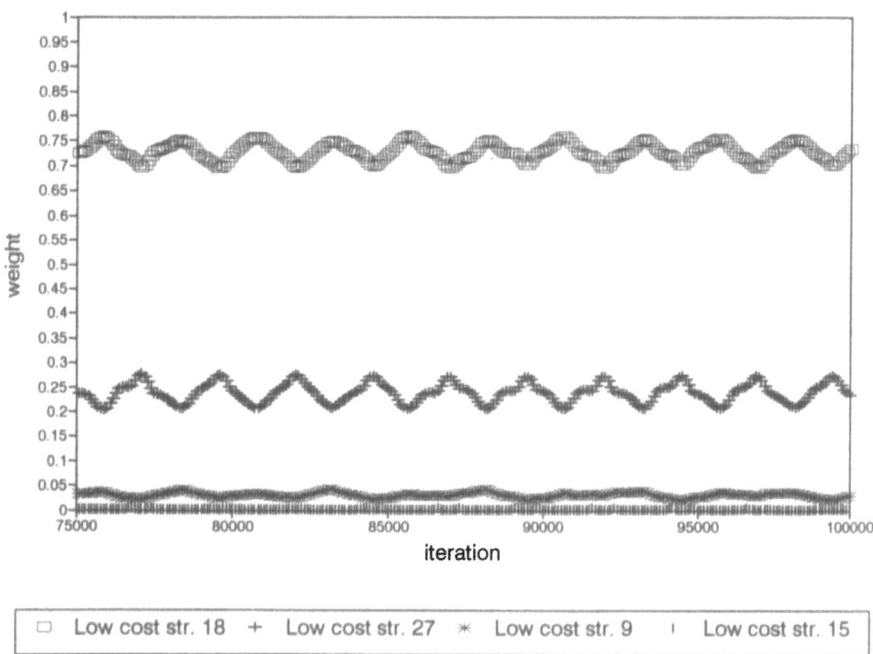

| □ Low cost str. 18 | + Low cost str. 27 | * Low cost str. 9 | ı Low cost str. 15 |

Figure 5.9 (Part II):   Weights of selected low cost strategies in an evolutionary process with 100 000 iterations

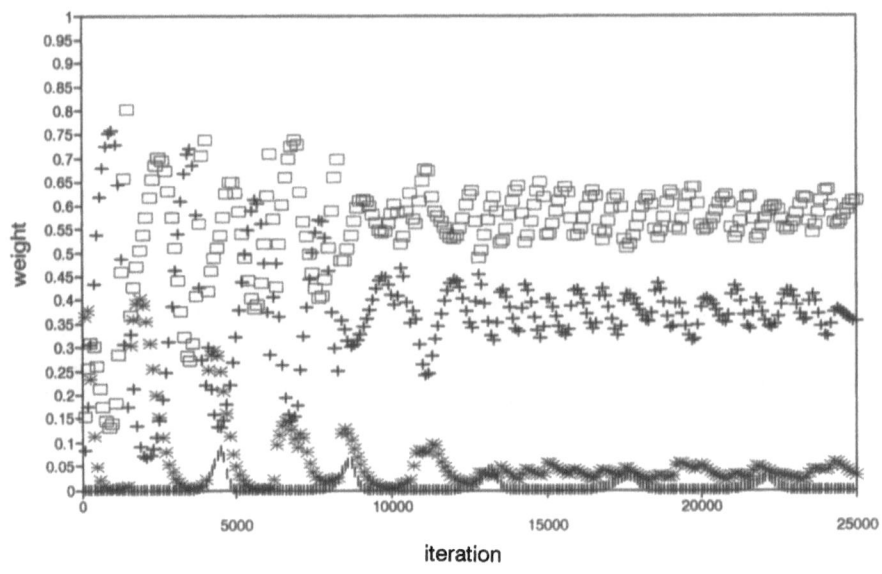

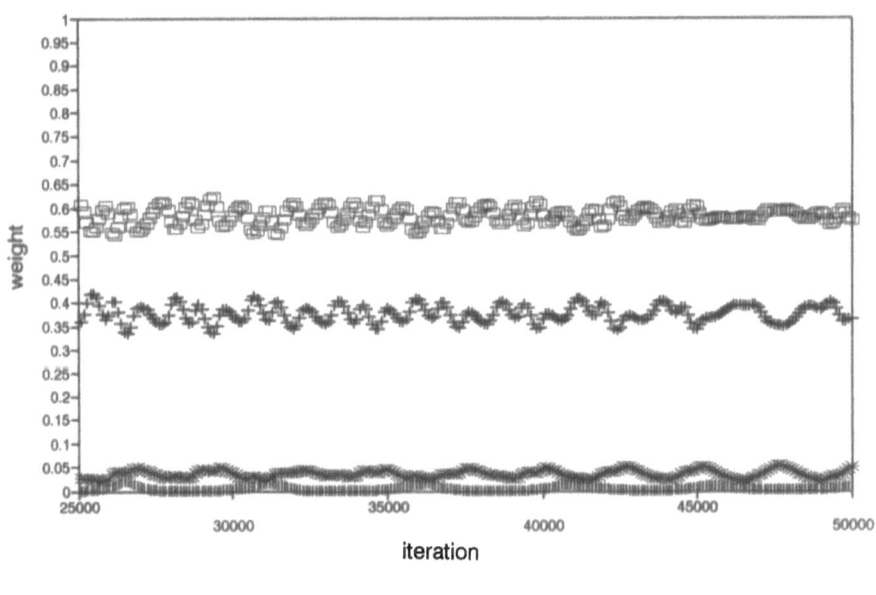

| □ High cost str. 17 | + High cost str. 31 | * High cost str. 10 | ı High cost str. 15 |

Figure 5.10 (Part I):   Weights of selected high cost strategies in an evolutionary process with 100 000 iterations

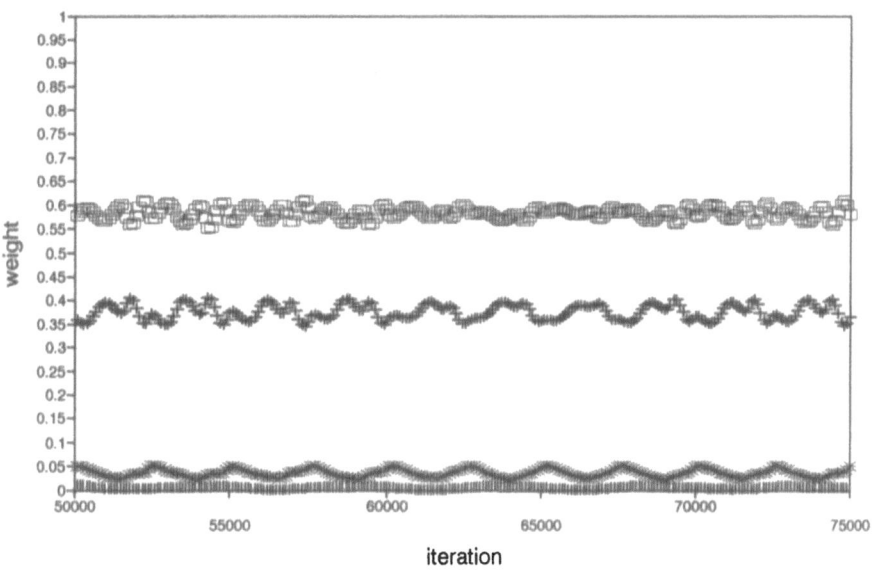

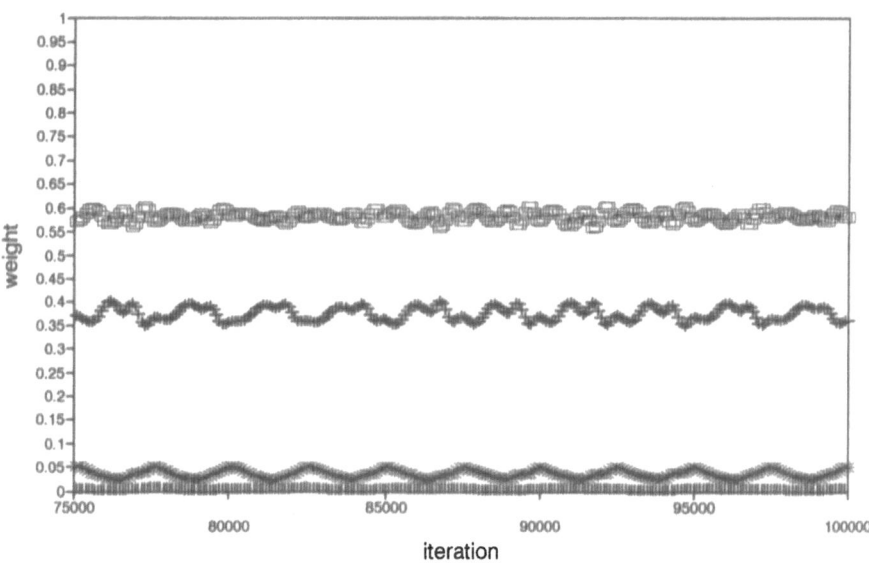

☐ High cost str. 17 + High cost str. 31 * High cost str. 10 ¦ High cost str. 15

Figure 5.10 (Part II):   Weights of selected high cost strategies in an
evolutionary process with 100 000 iterations

Table 5.28

Strategies' average weights in an evolutionary process with 100 000 iterations

participation no	low cost strategy	high cost strategy
1	.0000080	.0000036
2	.0000049	.0000024
3	.0000126	.0001104
4	.0000055	.0000056
5	.0021443	.0000665
6	.0000283	.0000201
7	.0000050	.0000227
8	.0000436	.0000614
9	.0260120	.0001239
10	.0065529	.0417136
11	.0000228	.0000258
12	.0000085	.0000018
13	.0000555	.0000483
14	.0000214	.0000098
15	.0018161	.0053801
16	.0000105	.0000160
17	.0000675	.5747454
18	.7224009	.0001142
19	.0000058	.0000057
20	.0000199	.0014742
21	.0000019	.0000003
22	.0000305	.0001018
23	.0000349	.0000745
24	.0001338	.0000038
25	.0000074	.0000116
26	.0000216	.0000160
27	.2403533	.0000259
28	.0000105	.0000160
29	.0000827	.0001977
30	.0000149	.0000410
31	.0000259	.3755411
32	.0000118	.0000036
33	.0000097	.0000093
34	.0000150	.0000059

Let us include low cost strategy no.15 into our equilibrium considerations as it makes an outstanding high profit against high cost strategy 17. We also need to include another high cost strategy; otherwise we would have more unknowns than equations if we want to compute the equilibrium probabilities of the low cost strategies. On the high cost side strategy no.10 imposes itself as a candidate because the time average of its weight of about 4% is the third highest on its cost side. Furthermore, low cost strategy no.15 makes a very low profit in the play against high cost strategy no.10. However, the three considered strategies on each cost side do not form an equilibrium of the 34x34 game. "Outsider" high cost strategy no.15 would be best a reply to the low cost strategy mixture where the probabilities are those of the completely mixed equilibrium of the 3x3 restricted game and all others zero.

We find an equilibrium, finally, if we extend the analysis to four strategies on each cost side. We include high cost strategy no.15 and, according to the time averages of the weights, low cost strategy no.9. The payoff matrix of the 4x4 restricted game and the probabilities with which the strategies are played in the completely mixed equilibrium of this game are presented in Table 5.29. This mixed equilibrium is an equilibrium point of the 34x34 game. On no cost side there is any pure strategy which makes a higher profit against the mixed equilibrium of the other cost side than do the four strategies of the restricted game. If we compare the theoretical probabilities of the strategies to the time averages of the strategies' weights we see that they come very close to each other.

Recall that in our evolutionary tournament we observe cyclical movements. Nevertheless, time averages closely agree with an equilibrium point of the 34x34 game. This result is not as surprising as it might seem at first glance. Hofbauer and Sigmund (1988, p.136) have shown that under certain conditions the time average of an orbit of the continuous replicator dynamic approaches a mixed equilibrium point in the limit. Even if the tournament is based on the discrete replicator dynamics modified by recurrent mutations it could be expected that something similar might happen in our case.

Table 5.29 (Part 1)

Payoff matrix of the 4x4 game where a low cost player can choose between the low cost strategies 9,15,18, and 27 and a high cost player can choose between the high cost strategies 10,15,17, and 31

high cost player

low cost player	H 10	H 15	H 17	H 31
L 9	152 203 99 960	180 835 86 541	157 525 98 636	154 334 99 324
L 15	73 878 109 326	169 974 66 400	178 341 46 804	129 680 74 933
L 18	154 830 102 650	160 459 103 355	156 039 102 970	156 652 103 301
L 27	154 509 105 481	157 798 105 401	154 521 105 092	159 071 103 807

Note: The low cost player's payoff is given in the upper left corner while the high cost player's payoff is given in the lower right corner

L is a strategy for the low cost situation

H is a strategy for the high cost situation

Table 5.29 (Part 2)

Probabilities in the mixed strategy equilibrium of the 4x4 game and
time average of the weights in the evolutionary process

strategy	equilibrium probability	average weight
L 9	.0318	.0260
L 15	.0016	.0018
L 18	.7280	.7224
L 27	.2386	.2404
H 10	.0359	.0417
H 15	.0049	.0054
H 17	.5831	.5747
H 31	.3761	.3755

Let me resume that we have found an equilibrium in mixed strategies of the
34x34 normal–form game which turns out to be near the time averages of the
evolutionary tournament. In this equilibrium point, four strategies on each cost
side are played with a probability larger than zero. However, only two strategies
on each cost side are played with a high probability. On each cost side one of
these is classified by us as cooperative while the other one is competitive.
However, the competitive strategies are only moderately aggressive: Low cost
strategy 27 is classified as competitive because it includes small price disturbances
in every fourth period. High cost strategy 17 is merely price oriented: It always
fixes a weighted average of the prices in the previous period. On the other hand,
the cooperative low cost strategy no.18 is at the borderline between cooperative
and competitive. We might consider it as being only very cautiously cooperative.
High cost strategy 31 is a clearly cooperative one which is in its reaction to
deviation more aggressive if the deviation was big than in the case that the
deviation was only small. Three of the other four strategies, which are played

with very small probabilities are strategies classified as competitive: low cost strategies 9 and 15 and high cost strategy 10. High cost strategy 15 is cooperative. Among them, low cost strategy 15 plays a crucial role although it has the lowest weight among the eight equilibrium strategies. This strategy starts playing with a seven–period–long extremely aggressive initial phase and then continues with a price of 130. It makes a very high profit when playing against high cost strategy 17, which is played with a considerable weight. However, it seems as if strategy no.15 itself cannot come to a really high weight in an evolutionary process as long as high cost strategy 10 has not completely died out. In the play of low cost strategy 15 against high cost strategy no.10, the latter makes for a high cost firm rather a high profit while low cost strategy no.15 makes only a very low one. Thus, if the weight of low cost strategy 15 increases in the evolutionary process, the weight of high cost strategy 10 increases as well, and as a consequence the weight of low cost strategy no.15 must decrease again. Low cost strategy 9 and high cost strategy 15 play only minor roles.

Note that the observations of our evolutionary tournament allow us some speculation about a long–run trend but only in a very restricted sense. The strategies on which the evolutionary tournament is based are those which have been submitted by subjects who had the experience of only one tournament round. We might suspect that the subjects would develop their strategies further if they played more rounds. Some subjects submitted for the second tournament strategies with a structure completely different from their strategies of the first tournament. In the evolutionary tournament, however, we restrict the strategy set to the strategies participating in the second tournament. New strategies cannot evolve during the evolutionary process. Ignoring this problem, we might consider our result as a hint where the trend might go to if we repeated our tournament more often: cautiously cooperative to moderately aggressive strategies. Unprotected cooperative and very aggressive strategies should disappear. The former are likely to disappear completely because they are exploitable. The latter might occur once in the while but they should not be able to spread out: They might, for example, provoke aggressive behavior on the opposite cost side which is more harmful to them than to the less aggressive strategies.

6. COMPARISON OF GAME–PLAYING EXPERIMENTS AND STRATEGY TOURNAMENTS

In this chapter we deal with the question to what extent the strategies also reveal the structure of subjects' spontaneous behavior in game–playing experiments. Let us start with a rough comparison of the outcomes of the game–playing experiments and the strategy tournaments.

In Chapter 5.1.1, Table 5.1, we have seen that on average over all plays realized long–run profits of each firm type are much higher in the tournaments than in both game–playing experiments.

In the game–playing experiments, average prices and profits were significantly higher in the second plays than in the first plays. In the second plays the subjects were more familiar with the game situation. Therfore, we want to restrict the following analysis mainly to the second plays of the game–playing experiments. Similarily, for reasons of subjects' experience, let us consider only the last tournament round.

Figure 6.1, for the low cost firms, and Figure 6.2, for the high cost firm, compare realized average prices in the second plays of the game–playing experiments to the prices realized on average in the second strategy tournament. As reference points, the figures also display the subgame perfect equilibrium prices and the prices in the myopic monopoly solution. We see that the prices fixed by the strategies are clearly higher than the average prices in the game–playing experiments.

We might suppose, however, that more repetitions of game–playing rounds by the same subjects would lead to similarly high prices. Prices increased significantly from the first to the second play. We suspect that they would increase further in subsequent plays. This speculation finds support in the following observation: Twelve of the sixty subjects participating in the game–playing experiments played a third and fourth round of the dynamic

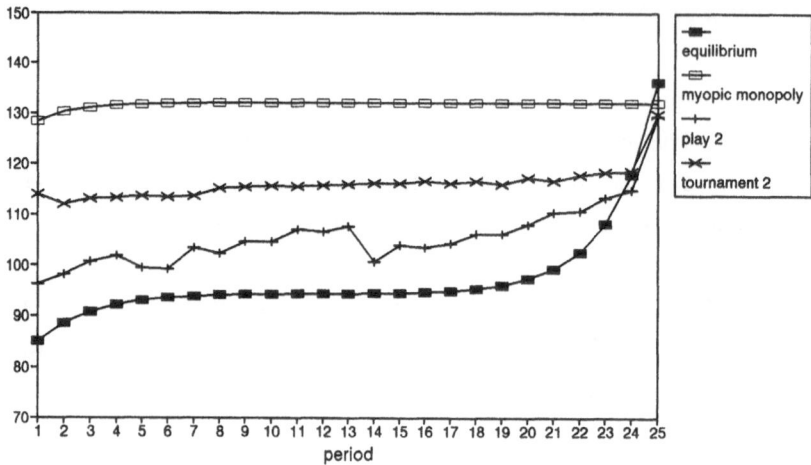

Figure 6.1:   LOW COST firms: Realized average prices in the second plays
of the game–playing experiments and in the second strategy
tournament, subgame perfect equilibrium prices and prices in
the myopic monopoly solution

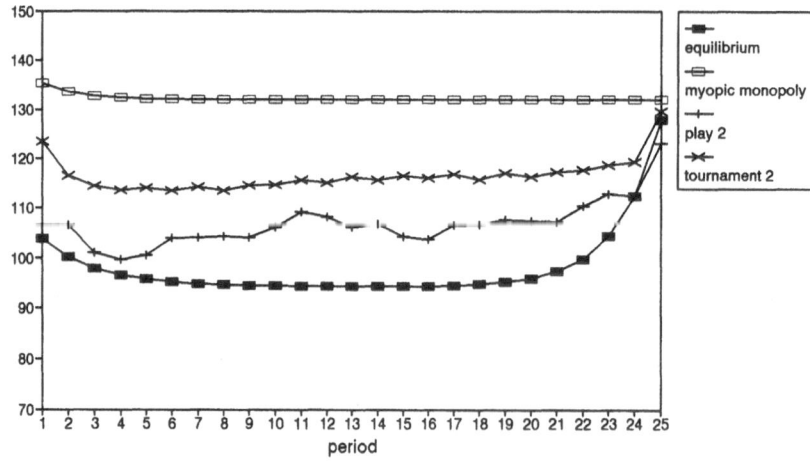

Figure 6.2:   HIGH COST firms: Realized average prices in the second plays
of the game–playing experiments and in the second strategy
tournament, subgame perfect equilibrium prices and prices in
the myopic monopoly solution

duopoly game in the same cost situation as before.[20] Let us consider the average realized prices in each period by each firm type in these plays. They are presented in Figures 6.3 and 6.4. As reference points, the figures also show the subgame perfect equilibrium prices and the prices in the myopic monopoly solution. We see that average prices are clearly higher than in the first two plays; they are rather close to the prices in the myopic monopoly situation. Note that we cannot recognize large differences in the average price levels in third and fourth plays. However, the average prices in the fourth plays seem more stable over time than the prices in the third plays. The average profits of low cost and high cost firms in third and fourth plays are given in Table 6.1. For both firm types, these profits are higher than the average profits in first and second plays. We conclude that more repetitions of game–playing rounds with experienced subjects are likely to lead to a more cooperative behavior, in the sense that higher prices are fixed which allow higher profits for both firm types.

In the game–playing experiments, we observe spontaneous behavior of subjects. In this situation, subjects do not deeply analyze the game. They do not consider a game analysis as their task. They take their decisions in a relatively short time even if no time limit is imposed; a deep analysis of the game would require some longer time. Contrary to this, in the strategy tournaments, we observe strategically planned behavior. Participants explore the game structure before they write a strategy. Therefore, we should expect in the strategically planned behavior more sophisticated concepts than in spontaneous behavior.

Let us analyze, for example, subjects' behavior in the last period in both experimental procedures. To fix one's actual myopic monopoly price is the only reasonable decision in the last period. Table 6.2 shows the percentages of subjects in the game–playing experiments and of strategies in the tournaments actually fixing the myopic monopoly price in the last period. In the game playing experiments we allow for deviations by less than one from the exact value. Note that price decisions in the game–playing experiments admitted two decimals only so that it might have been impossible to fix the exact value of the myopic monopoly price. Furthermore, by this we take also the possibility into account

---

[20] The results of these plays are not analyzed because they yield only three independent observations. This would not be enough for statistical purposes.

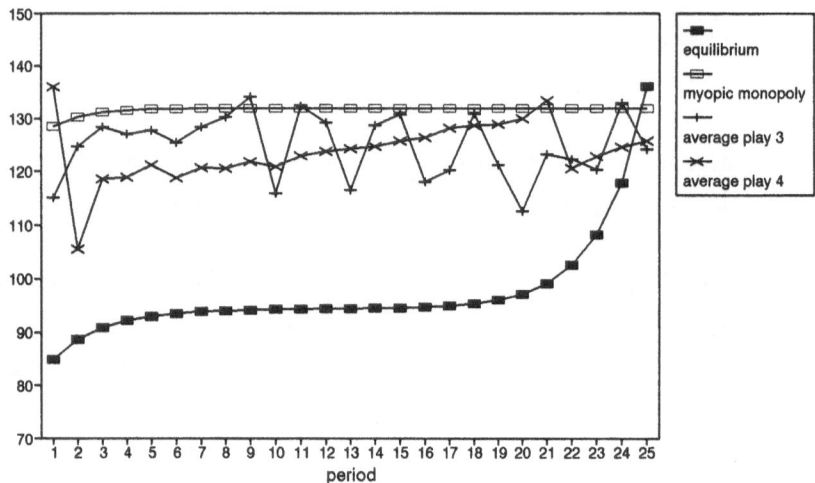

Figure 6.3: LOW COST firms: Realized average prices in third and fourth plays of game–playing experiments, subgame perfect equilibrium prices and prices in the myopic monopoly solution

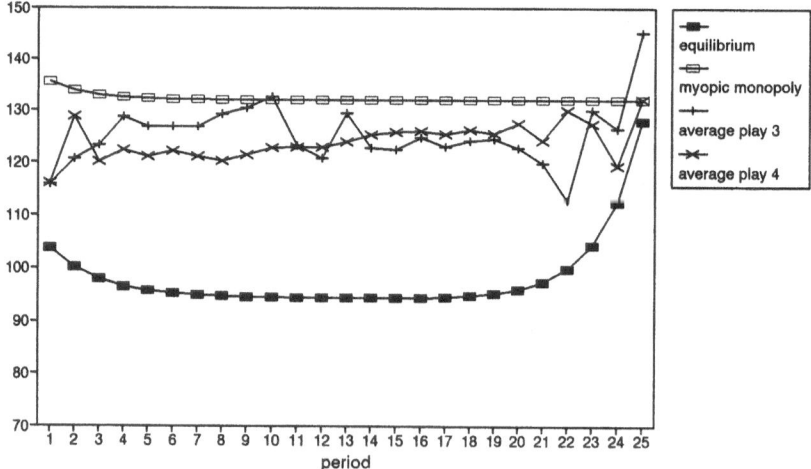

Figure 6.4: HIGH COST firms: Realized average prices in third and fourth plays of game–playing experiments, subgame perfect equilibrium prices and prices in the myopic monopoly solution

Table 6.1

Average realized long–run profits in third and fourth plays

	average long–run profit in	
firm type	3rd play	4th play
low cost	126 567	143 661
high cost	112 885	107 130

Table 6.2

Percentage of subjects in game–playing experiments and of strategies in tournaments fixing the myopic monopoly price in the last period

	game–playing experiments		strategy tournaments	
firm type	1st play	2nd play	1st tournament	2nd tournament
low cost	33	43	69	82
high cost	50	57	80	88

that subjects prefer integer numbers.[21] We see in Table 6.2 that only 50% of the subjects in the second plays of the game–playing experiments fix the myopic monopoly price in the last period while 85% of the strategies in the second tournament prescribe myopic monopoly pricing in the last period.

Note that neither in the game–playing experiments nor in the strategy tournaments we can find that setting the myopic monopoly price in the last period implies being more successful. However, if we consider the markets of the game–playing experiments where both firms fix the myopic monopoly price in the last period, we can show with a Mann–Whitney–U test that they make significantly higher market profits than the other markets. For both plays, the significance level of the one–tailed test is 2.5 %. Average ranks clearly indicate that these markets do better and not worse.

As another hint for strategic behavior being more sophisticated than spontaneous behavior, let us recall that about half of the strategies show a very similar feature: They involve a notion of cooperation. Typically, they consider myopic monopoly pricing as cooperative. These strategies aim at a partition of demand potential into 207 for the low cost firm and 193 for the high cost firm. Many strategies explicitly check in each period if their demand potential is at the required level of 207 or 193. Obviously, this typical feature of the strategies results from a detailed analysis of the game situation. In the game–playing experiments, we cannot clearly recognize what motivates the actual decisions. It seems as if demand potential also plays an important role for fairness considerations, but at another focal point that does not require deep reflections on the game: We suppose that in the game–playing experiments subjects typically want an equal share of demand potential. Comments by some subjects hint at this. Furthermore, in several markets we observe a frequent fight for demand potential. This fight finds expression in advantages of demand potential shifting from one firm to the other. It seems as if a firm's demand potential above 200, or in other words higher than the opponent's demand potential, is not accepted by the opponent firm which reacts then with a price reduction because it wants to retrieve its initial level of demand potential.

---

[21] The idea of prominence effects on human behavior was first mentioned by Schelling (1960).

Let me point to another difference between the behavior observed in game–playing experiments and the behavioral rules of the strategies. In the game–playing experiments, we sometimes observe abrupt price cuttings by a player. These abrupt price cuttings seem to be aimed at surprising the opponent. They are typical for markets classified as weakly cooperative.[22] A subject might abruptly reduce his price in order to increase his demand potential for the following period. He then increases his price above the initial level and makes a relatively high profit in this period following the price cutting. His calculation is that he should end up with more or less the same demand potential as before this "operation super profit". Strategies typically do not come up with price disturbances of this kind. Maybe, this is due to the fact that in game–playing experiments short–run adjustments to the opponent are possible. A subject who designs a strategy faces the disadvantage that he does not know what sort of opponents he is going to meet. He does not know how the opponents would react to such price disturbances. Therefore, we suspect that a strategy tends to play "safe" in the sense that it avoids to irritate its opponents. Opponents should be able to see through its motives.[23]

We know that strategies generally make their decision rules dependent on past prices and the actual demand potential. Other variables as, for example, short– or long–run profits are not taken into account. Let us give reasons for this: In a given period, the actual demand potential plays a crucial role. It is part of the demand function and, thus, influences strongly the profit possibilities in the considered and, to some degree, also the subsequent periods. It would not be sensible to form a profit aspiration without considering the actual demand potential. Furthermore, past price orientation is important because there is an interest in not deviating too much from the price level in the previous period(s). On the one hand, too high a price might lead to a loss in demand potential; on the other hand, too low a price might provoke aggressive opponent behavior. The two elementary variables, price and demand potential, are sufficient to determine most other variables which might be of interst. Other variables, as for example profits, need not be explicitly taken into consideration. Furthermore, they require a larger computational effort. Therefore, we suspect that also in the spontaneous

---

[22] See Chapter 4.2.

[23] Axelrod (1984) claims that in the repeated prisoner's dilemma situation such a feature is important for being successful in a strategy tournament. See also Chapter 7.

behavior the elementary variables, past prices and demand potential, play the most important role.

We suppose that there is a large conformity in the qualitative goals formed by subjects in the game–playing situation and when they design a strategy. In both situations, many subjects aim at a cooperation which finds expression in a fair partition of demand potential. We have seen, however, that the quantitative elaboration of this qualitative goal is different in spontaneous and strategic behavior. While we have evidence that strategic behavior often aims at a partition of demand potential at 207 and 193, we suppose that in the spontaneous behavior an equal share of demand potential is considered as fair by many subjects.

As it is very difficult to compare the strategies directly with the observed markets of the game–playing experiments, let us simulate game–playing experiments with the strategies. For one simulated experiment, we randomly match thirty low cost strategies each with a high cost strategy so that they form 30 duopoly markets. We do not permit that a strategy plays against a strategy of the opposite cost type submitted by the same participant. Nor is a strategy allowed to play in more than one market of the same simulated experiment. Let us consider a hundredfold repetition of such simulated experiments. Now, we may compare observed market behavior of the second plays of the game–playing experiments to simulated market behavior with the strategies.

Let us characterize the individual markets by some measures:

- market profit: the sum of profits of both firms in the market

- price level: average of the market prices over all 25 periods

- profit difference: long–run profit of low cost firm minus long–run profit of high cost firm

- demand potential difference: average demand potential over all 25 periods of the low cost firm minus average demand potential of the high cost firm

– price instability:	sum of the price instability measure given by equation (22) in Chapter 4.1.3 added up for the two firms in the market

To see if there are interdependencies between these measures characterizing a market, we compute rank correlation coefficients. Table 6.3 reports Spearman rank correlation coefficients for the second plays of the game playing experiments and the average rank correlation coefficients over the hundred simulation experiments. Coefficients with an absolute value of about .5 are significant on a 1% significance level. In the following, we call them "highly significant" correlations.

Figures 6.5 and 6.6 are graphical representations of the highly significant correlations in the game–playing experiments and the simulated experiments. Ignore, first, the arrow heads. In the simulated experiments, we observe first of all that there are less highly significant correlations than in the game–playing experiments. In the simulated experiments, price level, price instability and market profit are connected with each other but none of these variables is connected to one of the two other variables, demand potential diffence and profit difference. The strongest correlation is between price level and market profit; it is close to one. Price instability is negatively correlated with price level and market profit. The strong correlation between price level and market profit can also be observed in the game–playing experiments. But, in the game–playing experiments there is no highly significant correlation between price level and price instability. However, we find here additional highly significant correlations: Profit difference and demand potential difference are very significantly negatively correlated with price level and market profit. In both game–playing experiments and simulated experiments, profit difference and demand potential difference are correlated with each other by almost one.

Table 6.3

Rank correlation coefficients (Spearman) in the game–playing experiments
and average rank correlation coefficients of 100 simulated experiments

	price level	price instability	profit difference	demand potential difference
market profit	.96 .93	−.48 −.69	−.49 .07	−.52 .08
	price level	−.35 −.63	−.51 −.08	−.53 −.04
		price instability	.34 .21	.24 .11
			profit difference	.91 .95

Note: The upper number in each box is the correlation coefficient in the game–
playing experiments; the lower number is the average correlation coefficient
in the 100 simulated experiments.

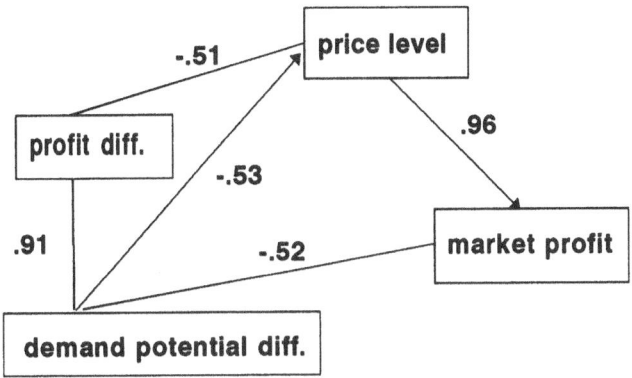

Figure 6.5: Rank correlation coefficients which are significant at
a 1% level in the game–playing experiments

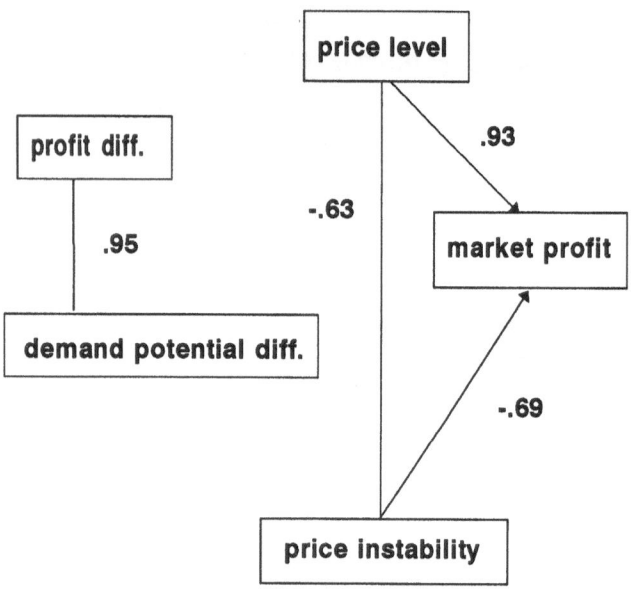

Figure 6.6: High average rank correlation coefficients in the 100 simulated experiments. Only those average rank correlation coefficients are included which would be highly significant in a single experiment.

Let us try to give a causal interpretation of these interdependencies. Doing this, we want to neglect the profit difference variable. It is correlated with the demand potential difference by almost one and we might consider this relationship as somehow tautological. We try to construct "causality chains" for the remaining measures. Let us start with the game–playing experiments. The demand potential difference is observed before a price decision is taken; this variable might be considered as influencing the price level. The price level then determines the market profit. This is the interpretation which seems most plausible by my intuition although others would be possible. Let us give a similar reasoning for the causalities in the simulated plays: Again, the price level has a very strong influence on the market profit. However, also the price instability has quite a strong influence. Price level and price instability are highly correlated with each other. It is difficult to attribute here a direction of causality. The demand potential difference is not connected to this causality chain. This might be due to the fact that there are no very aggressive low cost strategies participating in the second tournament which try to force the opponent out of the market by setting permanently very low prices. The directions of causality resulting from these interpretations are indicated by the arrow heads in Figures 6.5 and 6.6.

Obviously, in the game–playing experiments there are other forces at work than in the simulated experiments. We want to test if the interdependencies between the considered variables altogether are different in both cases. To compare the complete system of interdependencies in game–playing experiments and in simulated experiments, we construct a heuristic "randomization test". For each of the hundred simulated experiments, we compute the sum of square deviations of the correlation coefficients from the corresponding average correlation coefficient over the hundred plays. Also for the game–playing experiments, we compute the sum of square deviations of the correlation coefficients from the corresponding average correlation coefficient of the hundred simulations. Let us call these measures the "correlation square deviations". The hundred correlation square deviation values of the simulated experiments describe the distribution of the correlation square deviation in the simulated experiments. We count how many correlation square deviations of the simulated experiments lie above the correlation square deviation of the game–playing experiments. This number gives us an estimate of the percentile of the distribution at which the value for the game–playing experiments is located. The correlation square deviation of the game–playing experiments is 1.29 while the highest correlation square deviation of the simulated plays is only .81. Thus, the correlation square deviation of the game–playing experiments is in the 0–percentile. A one–tailed Mann–Whitney–U test rejects the null hypothesis on the 5% significance level. We conclude that the correlation structure of the game–playing experiments is different from the correlation structure in the simulated experiments. We interpret this result as statistical evidence that there are somewhat different quantitative goals driving the strategic and the spontaneous behavior.

Let us resume that behavior in game–playing experiments does not seem as systematic as strategy behavior. We do not get a clear picture what really determines spontaneous behavior but we suppose that the price decisions result in both spontaneous and strategic behavior from a similar qualitative thinking. There is evidence, however, that the quantitative transformation of the qualitative goals is different in strategies and spontaneous behavior. This is mainly due to a sophisticated analysis of the game situation in case that a strategy is to be designed while subjects in the game–playing situation do not analyze the game in similar depth. There is, however, some evidence that more experience in spontaneous playing of a given game leads to a behavior which is more similar to the strategy behavior.

7.   COMPARISON WITH RELATED STUDIES

In the first three sections of this chapter, I describe very briefly three other experimental studies. In Section 4 the findings of these studies will be related with each other and with the results of my study of experimental duopoly markets with demand inertia. Therefore, I restrict the presentation of these studies and their results essentially to the aspects which are relevant for the discussion in Section 4.

7.1   Selten (1967a,b)

Selten (1967a,b) experimentally analyzed a price setting oligopoly with investment. In his model, demand is characterized by demand inertia. Selten (1967a) describes the results of game–playing experiments while Selten (1967b) introduces the strategy method as a method of experimentation: Subjects, after having acquired experience in playing the duopoly game, are asked to design a strategy for this game in flow–chart–form. The structure of these strategies is then analyzed.[24]

The game situation of Selten's experiments is similar to the game situation analyzed in my study: Each of three firms, at the end of each of thirty periods, has to decide upon its price. It has also to decide if it wants to increase or reduce its capacity. It knows all past prices of the opponents. It does, however, not know the production costs, past sales and the assets of the opponents. It also has incomplete information about the demand function. It knows that the higher the price the lower will be the demand. Furthermore, it knows that there is demand inertia: Customers tend to buy from whom they used to buy; but they also have a tendency to change from high price firms to low price firms. The total demand increases over time.

---

[24] For technical reasons, it was at this time hardly possible to organize computer tournaments where these strategies could have competed with each other. Tietz (1967), however, made a simulation study with only a small number of strategies which he had designed following Selten's results.

There is no theoretical solution for this game situation. As an approach, however, Selten considers a fictitious situation, based on the assumption of perfect competition, and presents a solution based on simplifying criteria. This solution describes quite well some features of the observed behavior in the game–playing experiments. (Selten (1967a))

The strategies designed by the subjects reveal the following fundamental ideas of the investment and price policy (Selten (1967b)):

If capacities are not fully utilized, subjects try to increase sales. Their aim is full utilization of capacities.

If capacities are fully utilized there are two possibilities:
- No investment: Then, sales cannot be increased any more; the price is fixed as high as possible such that it does not endanger the utilization of capacities.
- Investment: Then, from a price policy point of view, the situation is similar to the situation when capacities are not fully utilized. In this situation it is of importance to increase sales.

Voluntary liquidation does not occur. It is costly. Investments are made only if the capacities have been fully utilized in the previous period.

Two additional investment criteria frequently appear in the strategies:
1) Investments should not lead to a negative balance.
2) The expected net profit should be higher with investments than without.

In the price policy of most strategies, the opponents' prices play an important role: The price situation determines the extent and the direction of a possible price change. Some of these strategies additionally check if the expected gross profit in case of a price change increases compared to the gross profit in the previous period or the gross profit to be expected in case of no price change. Only in case of an increase the price change is effectuated.

Neither in the game–playing experiments nor in the strategies cooperative behavior is observed.

7.2 Axelrod (1984)

A famous application of the strategy method, organized as computer tournaments with a general invitation to participate, are Axelrod's (1984) tournaments for a finitely repeated prisoner's dilemma game.

The prisoner's dilemma game may be presented as a two–person normal–form game where both players have to chose between two pure strategies: COOPERATE and DEFECT. If both players COOPERATE then both get a payoff, R, higher than their payoff, P, in the case that both DEFECT. However, it is a dominant strategy for both players to DEFECT. If one player DEFECTS unilaterally he gets a payoff, T, which is higher than R while the other player's payoff, S, is lower than P. The payoffs also satisfy R > (T+S)/2. Thus, the equilibrium of this game is, for both players, to chose DEFECT. It leads to a Pareto–inefficient outcome. In the repeated prisoner's dilemma each move is worth less than the move before, by a given discount factor.

A very simple strategy called "tit–for–tat" won Axelrod's tournaments. Tit–for–tat is a strategy which plays cooperative in the first period and then always imitates the opponent's behavior in the previous period. Note that this strategy, when playing once the repeated prisoner's dilemma game, never makes a higher profit than its opponent. The reason why tit–for–tat nevertheless is successful in the tournaments is that it does well in the interactions with a wide variety of strategies while other strategies may have difficulties to lead opponent strategies towards cooperation. Axelrod worked out four characteristics which are important for being successful in a repeated prisoner's dilemma tournament:

- Be nice: Don't be the first to deviate from cooperation.
- Reciprocate both cooperation and defection.
- Don't be envious: If the opponent returns, after a phase of defection, to cooperation then forgive him and also cooperate again.
- Don't be too clever: Don't keep your intentions hidden.

Tit–for–tat satisfies all these requirements.

7.3 Selten, Mitzkewitz and Uhlich (1988)

Selten, Mitzkewitz and Uhlich (1988) applied the strategy method by organizing a student seminar. Participants played three rounds of game–playing experiments of a twenty–times–repeated asymmetric Cournot duopoly with linear costs and demand. Then, there followed three rounds of strategy programming for the numerically specified game. Each round ended with a computer tournament where the strategies were played against each other. After the first two rounds, the participants had the opportunity to improve their strategies in light of the tournament results. The success in the last tournament determined a student's grade for the seminar.

The strategies of the last tournament reveal the following typical structure: A distinction is made between initial phase, main phase, and end phase. The initial phase serves to signal the willingness to cooperate. In the main phase, the strategy tries to maintain cooperation. In the end phase, cooperation is given up. The typical behaviour of the main phase follows a "measure–for–measure" policy. This policy aims at cooperation at an "ideal point". The ideal point describes the player's own cooperative goal. He does not know the cooperative goal of the other player. The measure–for–measure policy responds to movements of the opponent towards the player's ideal point or away from it by similar movements in the same direction.

Selten, Mitzewitz and Uhlich measure the typicity of a strategy by an index of typicity based on thirteen characteristics. They find a high positive correlation between this index and the success of a strategy in the final tournament.

7.4 Comparison of results

To begin with, let us compare the results by Selten, Mitzkewitz and Uhlich (1988) and Axelrod (1984). The tit–for–tat strategy which was the winner of Axelrod's tournaments for the repeated prisoner's dilemma is fully in harmony with the measure–for–measure principle which was typical for the strategies in the repeated Cournot duopoly tournament. In the prisoner's dilemma game, there is no question where cooperation should take place, and as there are only two choices, measure–for– measure cannot mean anything else but tit–for–tat. Thus, the tit–for–tat principle may be considered as a translation of the (more general) measure–for– measure principle to the prisoner's dilemma situation.

The translation of the tit–for–tat or the measure–for–measure principle to the dynamic duopoly situation is less clear. The most straightforward translation of the tit–for–tat principle into the dynamic duopoly situation would be price imitation. Although we do not observe, in our tournaments, pure price imitation, there are several strategies which are past price oriented but consider additional goals. We observe that many of these past price oriented strategies do not start with a cooperative price.[25]

Tit–for–tat does not explicitly require the definition of a cooperative goal because cooperation is so obvious in the prisoner's dilemma game. The more general measure–for–measure principle is more relevant than tit–for–tat in the dynamic dupoly situation. Measure–for–measure is based on the formation of a cooperative goal. The strategies which are classified by us as cooperative behave essentially according to this principle. We require that a cooperative strategy involves a notion of cooperation, a cooperative goal, and that it reacts to deviations from this goal. Furthermore, we require cooperative behavior in the beginning.[26] We may consider about half of the strategies as applying the measure–for–measure–principle.

Note that price imitating behavior has also been observed very often by Selten (1967). Definition of cooperative goals, however, did not occur in his strategies.

---

[25] 70% of the past price oriented low cost strategies and 50% of the past price oriented high cost strategies start with a price below 128.50.

[26] Note that we consider as cooperative also unified strategies which involve a notion of cooperation but do not react to deviation from cooperation. As an example, consider the strategy which always sets its myopic monopoly price. In a sense, this strategy also reacts to deviation but only in a very moderate way: If the opponent in the previous period has fixed a lower price, the strategy's demand potential is reduced so that the new monopoly price is lower.

In contrast to Selten, Mitzkewitz and Uhlich, the cooperative strategies in our tournaments are not likely to be the most successful. This might be due to conflicting goals in the duopoly game with demand inertia: There is a myopic goal of making a high profit in an actual period while there is also a long–run goal of protecting one's demand potential. The importance of guarding the demand potential leads to an interest in being a little bit aggressive. One should, however, be only so little aggressive that one's own aggressiveness does not provoke aggressive reactions of the cooperative opponents although one exploits to some extent their cooperative behavior. A strategy which is moderately aggressive is likely to be among the winners of a tournament. It risks, however, to exceed the limit of tolerated aggressiveness and, therefore, to be very unsuccessful. Cooperation with reaction to deviation, or the application of the measure–for–measure principle, leads to a safe medium success in our tournaments.

We observe in our strategies a phase structure distinguishing between initial phase, main phase, and endphase. Selten, Mitzkewitz and Uhlich (1988) observe a similar phase structure. However, there are different motives for the distinction of these phases. In Selten, Mitzkewitz and Uhlich (1988), the initial phase typically serves to signal the willingness of cooperation. In the strategies of our tournaments, the specification of an initial phase seems to result simply from the necessity to fix a starting price. Similarily, a different behavior rule for the last period is needed in most strategies for the duopoly with demand inertia because myopic monopoly pricing is the only reasonable behavior in the last period. A breakdown of cooperation in the end phase, as observed by Selten, Mitzkewitz and Uhlich, is not typical in the strategies of our study.

Both studies, Selten, Mitzkewitz and Uhlich and our study have in common that no predictions about opponent's behavior are made.

In contrast to the other studies discussed here, cooperation does not appear in the strategies of Selten (1967b). Let me give two possible reasons for cooperation being less likely in the game situation analyzed by Selten than in the other game situations. First, it is not clear where cooperation should take place in the game analyzed in Selten's experiments. Secondly, there are "rigidities": Demand inertia makes long–run considerations important. A low, non cooperative initial price is likely to increase one's future demand potential. This increase is then effective for the rest of the game. There is an additional, even stronger, rigidity in Selten's model, caused by the possibility of investment. Desinvestment is costly. Therefore, once an investment is made, the subject's major goal is to utilize the capacity. Reduction of capacity is not taken into consideration. Such rigidities create accomplished facts. They give the possibility to build up long–run advantageous positions by a deviation from cooperation but make it difficult to achieve cooperation in later periods.[27]

These reflections lead me to the following hypotheses:

1) The more obvious it is in a game situation where cooperation should take place the more likely it is to influence human behavior.

2) "Rigidities" (as demand inertia and investment) make cooperation less likely because they give the possibility to create accomplished facts in one's own favor.

To support these hypotheses, let us reconsider briefly the four studies. In Axelrod (1984) and Selten, Mitzkewitz and Uhlich (1988) the possibility of cooperation is obvious. There are no rigidities: In each period the players are in the same game situation whatever the past history. In both studies, we observe that cooperation with reaction to deviation (tit for tat, measure–for measure) is typical and successful. In our study, the possibility of cooperation is obvious but there is some rigidity caused by demand inertia. We observe that cooperation is typical but it is not most successful in the tournament. A moderate aggressiveness makes a strategy more successful. In Selten (1967a,b), it is not obvious where cooperation should take place and there is strong rigidity caused by investment and demand inertia. Cooperation is not observed.

---

[27] Another argument for less cooperation in Selten's experiments might be that total demand is increasing over time. The demand increase favors early investments.

Of course, these hypotheses would need to be tested systematically in experiments. In order to test the second hypothesis, we might examine experimental oligopoly markets where investment is a decision variable. Costs of liquidation (desinvestment) could be considered as describing the rigidity caused by the possibility of investment. In experiments, we might systematically vary these costs and, by this, the degree of rigidity.

Another interesting question to be analyzed in further experiments concerns oligopoly games with incomplete information. Will we observe cooperative behavior in games with incomplete information? All considered studies, with the exception of Selten (1967a,b), examine games with complete information. The incompleteness of information might be another reason why cooperation is not observed by Selten (1967a,b).

8.   SUMMARY

Results of the game–playing experiments

The actual behavior of subjects is different from what is prescibed by the subgame perfect equilibrium solution of the game.

There occurs significantly more cooperative behavior in second plays than in first plays: Profits, prices, and stability of prices are higher.

We distinguish between three market types:

−	Strongly cooperative markets:	Both firms set permanently high prices
−	Weakly cooperative markets:	Prices fluctuate a lot in a middle price range.
−	Aggressive markets:	One firm tries to force the other out of the market by sticking to extremely low prices.

These market types can also be described by the average market price level and the price instability of the two firms in the market:

- Cooperative markets show a higher price level than aggressive markets.
- Weakly cooperative markets show a higher price instability than strongly cooperative markets.

All observed markets of the second plays are captured by this clustering. Among 30 markets, we identify 9 strongly cooperative markets, 10 weakly cooperative markets and 11 aggressive markets.

Some of the strongly cooperative markets realize profits very close to the Pareto frontier.

Results of the strategy tournaments

Average profits in both tournaments are clearly higher than the profits in the game–playing experiments.

Average profits are almost the same in both tournaments although in the first tournament there is a higher variation of profits than in the second tournament.

In both tournaments, we observe many markets that realize profits close to the profits in the case that both firms always fix their myopic monopoly prices.

Almost half of the strategies have a similar feature: They involve a notion of cooperation which is typically myopic monopoly pricing. In each period, they check if the opponent has behaved cooperatively in the previous period. If this is the case they also cooperate, otherwise, they react to the deviation. In the first period they start with a cooperative price. We classify these strategies as cooperative.

All strategies with a different structure are classified as competitive. Among the strategies classified as competitive, there are many strategies imitating the opponent's price in the previous period. Competitive strategies typically start with a price below the myopic monopoly price of the low cost firm.

Competitive strategies are more aggressive than cooperative strategies, in the sense that their opponents make lower profits.

The dispersion of average profits of the cooperative strategies is lower than the dispersion of average profits of the competitive strategies: Cooperative strategies are more likely to end up in the middle part than at the top or the bottom of the ranking list of the tournament success. Competitive strategies are more likely to be either very successful or very unsuccessful.

Competitive strategies are very successful in the tournaments if they are only slightly aggressive so that they can exploit the cooperative strategies without causing aggressive reactions. Competitive strategies risk, however being very unsuccessful by being overly aggressive and therefore provoking aggressive opponent behavior.

The strategies reveal a phase structure: Typically, there is an initial phase to fix a starting value. An endphase serves to fix the myopic monopoly price in the last period. 80% of the strategies in the second tournament actually fix the myopic monopoly price in the last period. In the main phase, decisions depend only on the elementary variables demand potential and past opponent prices. The complexity of recall is one period. No predictions are made of the opponent's behavior. 40% of the strategies in the second tournament are observed to set occasionally prices above their actual myopic monopoly prices.

The structures of low and high cost stategies are very similar although the low cost strategies sometimes behave more assertively than the high cost strategies, in the sense that they react in a sharper way to deviation from cooperation.

After the first tournament, there was a slight tendency towards better protection of the low cost strategies against aggressive opponent behavior while the high cost strategies tended to show a somewhat more aggressive behavior. This effect is, however, very weak because about one third of the strategies was not changed at all.

An evolutionary tournament with the strategies of the second tournament

We consider an evolutionary process with 100 000 iterations based on the discrete replicator dynamics modified by recurrent mutations. We observe that two high cost and two low cost strategies survive with high "weights". Their weights show cyclical movements which convey the impression that they are approaching limit cycles.

In the 34x34 normal–form game where a low cost player and a high cost player can choose among the second tournament strategies, we find a mixed equilibrium with positive probabilities for the four main surviving strategies and two other strategies on each cost side. The equilibrium probabilities come very close to the observed time averages of the strategies' weights. The strategies which survive with a high weight can be described as cautiously cooperative or moderately aggressive.

Comparison of game–playing experiments and strategy tournaments

Comparing the results of both experimental methods, we find support for the hypothesis that strategic behavior is based on more sophisticated concepts than spontaneous behavior in the game–playing experiments. The concepts on which the strategies are based result obviously from a deeper analysis of the game. In the game–playing situation the game is analyzed only superficially. For example, the strategies typically aim at the myopic monopoly solution while spontaneous behavior seems to consider the equal share of demand potential as a fair situation. Spontaneous behavior allows short–run adjustments to the opponent so that we sometimes observe, for example, abrupt price movements in spontaneous behavior. Strategic behavior, however, is more systematic and avoids irritation of the opponent.

To make the spontaneous behavior and the behavior induced by strategies directly comparable, we simulate game–playing experiments with the strategies. A rank correlation analysis between some measures characterizing markets leads us to the conclusion that the correlation structure as a whole is different in game–playing experiments and simulated experiments.

Although spontaneous behavior and strategic behavior seem to be guided by similar qualitative goals, the transformation into quantitative goals is clearly different.

Comparison with related studies

The strategies classified by us as cooperative are in harmony with the measure–for–measure principle by Selten, Mitzkewitz and Uhlich (1988).

However, while in the study by Selten, Mitzkewitz and Uhlich (1988) the measure–for–measure principle is typical and successful, somewhat aggressive strategies are more successful than the typical cooperative strategies of this study.

Considering also the results by Selten (1967a,b) I formulate the hypothesis that "rigidities" in the oligopoly situation, caused for example by investments and demand inertia, make cooperative behavior less likely because they present the opportunity to gain long–run advantages by a deviation from cooperation.

REFERENCES

ALGER, D. (1986) : Investigating oligopolies within the laboratory. Bureau of
Economics, Federal Trade Commission, Washington.

AXELROD, R. (1984) : The evolution of cooperation. Basic, New York.

BENSON, B.L. / FAMINOW, M.D. (1988) : The impact of experience on prices
and profits in experimental duopoly markets. Journal of Economic
Behavior and Organization 9, 345–365.

BOURDIEU, J. / SERVAIN, M. (1991) : DELTA au tournoi I.D.E.A.S.. La
Lettre Δ, Paris.

FRIEDMAN, J.W. / HOGGATT, A.C. (1980) : An experiment in noncooperative
oligopoly. Research in experimental economics, Vol.1, Supplement 1,
JAI Press,Greenwich.

HOFBAUER, J. / SIGMUND, K. (1988) : The theory of evolution and dynamical
systems. Cambridge University Press, Cambridge.

KESER, C. (1989) : IDEAS – International Duopoly Experiments And
Simulations : Call for strategies. Mimeo, University of Bonn.

LIENERT, G.A. (1973) : Verteilungsfreie Methoden in der Biostatistik, Band I.
Verlag Anton Hain, Meisenheim am Glan.

LIENERT, G.A. (1975) : Verteilungsfreie Methoden in der Biostatistik,
Tafelband. Verlag Anton Hain, Meisenheim am Glan

MAYNARD SMITH, J. (1982) : Evolution and the Theory of Games.
Cambridge University Press, Cambridge.

ROUGHGARDEN, J.(1979) : Theory of population genetics and evolutionary ecology. MacMillan, New York.

SCHELLING, T.C. (1960) : The strategy of conflict. Harvard University Press, Cambridge, Massachusetts, and London, England.

SELTEN, R. (1965) : Spieltheoretische Behandlung eines Oligopolmodells mit Nachfrageträgheit. Zeitschrift für die Gesamte Staatswissenschaft, Bd.121, 301–324 and 667–689.

SELTEN, R. (1967a) : Ein Oligopolexperiment mit Preisvariation und Investition. In: Sauermann (ed.), Beiträge zur experimentellen Wirtschaftsforschung, J.C.B. Mohr, Tübingen, 103–135.

SELTEN, R. (1967b) : Die Strategiemethode zur Erforschung des eingeschränkt rationalen Verhaltens im Rahmen eines Oligopolexperiments. In: Sauermann (ed.), Beiträge zur experimentellen Wirtschaftsforschung, J.C.B. Mohr, Tübingen, 136–168.

SELTEN, R. (1990) : Bounded rationality. Journal of Institutional and Theoretical Economics, 146, 649–658.

SELTEN, R. / MITZKEWITZ, M. / UHLICH, G.R. (1988) : Duopoly strategies programmed by experienced players. Discussion Paper No. B–106, University of Bonn.

SIEGEL, S. (1957) : Nonparametric statistics for the behavioral sciences. New York–Toronto–Tokyo: McGraw–Hill.

STOECKER, R. (1980) : Experimentelle Untersuchung des Entscheidungsverhaltens im Bertrand–Oligopol. Pfeffer, Bielefeld.

TIETZ, R. (1967) : Simulation eingeschränkt rationaler Investitionsstrategien in einer dynamischen Oligopolsituation. In: Sauermann (ed.), Beiträge zur experimentellen Wirtschaftsforschung, J.C.B. Mohr, Tübingen, 169–225.

APPENDIX A

List of Participants

1.   **Michael Badke**, Universität Bonn, Germany
2.   **Jürgen Bartnick**, Universität Frankfurt, Germany
3.   **Pierre Dehez**, European University Institute, Firenze, Italy
4.   **LS–Mikroökonomik Uni Dortmund**, Universität Dortmund, Germany
5.   **3 x III**, Universität Bonn, Germany
6.   **Klaus Fischer**, München, Germany
7.   **Hans Haller**, Virginia Polytechnic Institute and State University, USA
8.   **Anonymous 1**, France
9.   **IFA Göttingen**, Universität Göttingen, Germany
10. **Die Glücksritter**, Universität der Bundeswehr München and Universität Bonn, Germany
11. **Tobias Klaus**, Universität Bonn, Germany
12. **KÖLNBNNS**, Universität Köln, Germany
13. **Manfred Königstein**, Universität Frankfurt, Germany
14. **Kai–Uwe Kühn**, Nuffield College, GB
15. **Anonymous 2**, GB
16. **Stephen Martin**, European University Institute, Firenze, Italy
17. **MA–Team**, Universität Mannheim, Germany
18. **Anonymous 3**, Germany
19. **Martin Siefen, Rüdiger Frey, and Alexander Stremme**, Universität Bonn, Germany
20. **Monika Schnitzer and Klaus Schmidt**, Universität Bonn, Germany
21. **Seppo Suominen**, Helsinki University of Technology, Finland
22. **Franz Waldenberger**, Universität Köln, Germany
23. **Klaus Wehrt**, Universität Hamburg, Germany
24. **Chun–Lei Yang**, Universität Dortmund, Germany
25. **University of Basle**, Universität Basel, Switzerland
26. **EUIDEAS**, European University Institute, Firenze, Italy
27. **Torsten Füg**, Universität Bonn, Germany
28. **Anonymous 4**, Germany

29. **LEBEC**, Universität Graz and Universität Heidelberg, Austria and Germany

30. **K.U. Leuven**, Katholieke Universiteit Leuven, Belgium

31. **Anonymous 5**, Norway

32. **NOVICE**, University of California, Los Angeles, USA

33. **Jan Potters, Arthur Schram, Hans van Ophem, Eric Drissen, and Frans van Winden**, Universiteit van Amsterdam, The Netherlands

34. **Die Drei von der Tankstelle**, Universität Bonn, Germany

35. **Philippe Jehiel**, Universität Bonn, Germany

36. **Julia Lambach and Achim Preuß Neudorf**, Universität Bonn, Germany

37. **Anonymous 6**, GB

38. **Ca'Bembo–Venezia**, Università di Venezia, Italy

39. **Helmut Bester**, Universität Bonn, Germany

40. **Astolfo/Chatfield (students of Myrna Wooders, University of Toronto)**, University of Toronto, Canada

41. **MERIT**, Maastricht Economic Research Institute on Innovation and Technology, The Netherlands

42. **Rosemarie Nagel**, Universität Bonn, Germany

43. **DETAL**, Ecole Normale Supérieure, Paris, France

44. **MARS**, Universität München, Germany

45. **Uni Konstanz**, Universität Konstanz, Germany

APPENDIX B

Examples of actually observed markets in the game–playing experiments

A STRONGLY COOPERATIVE MARKET

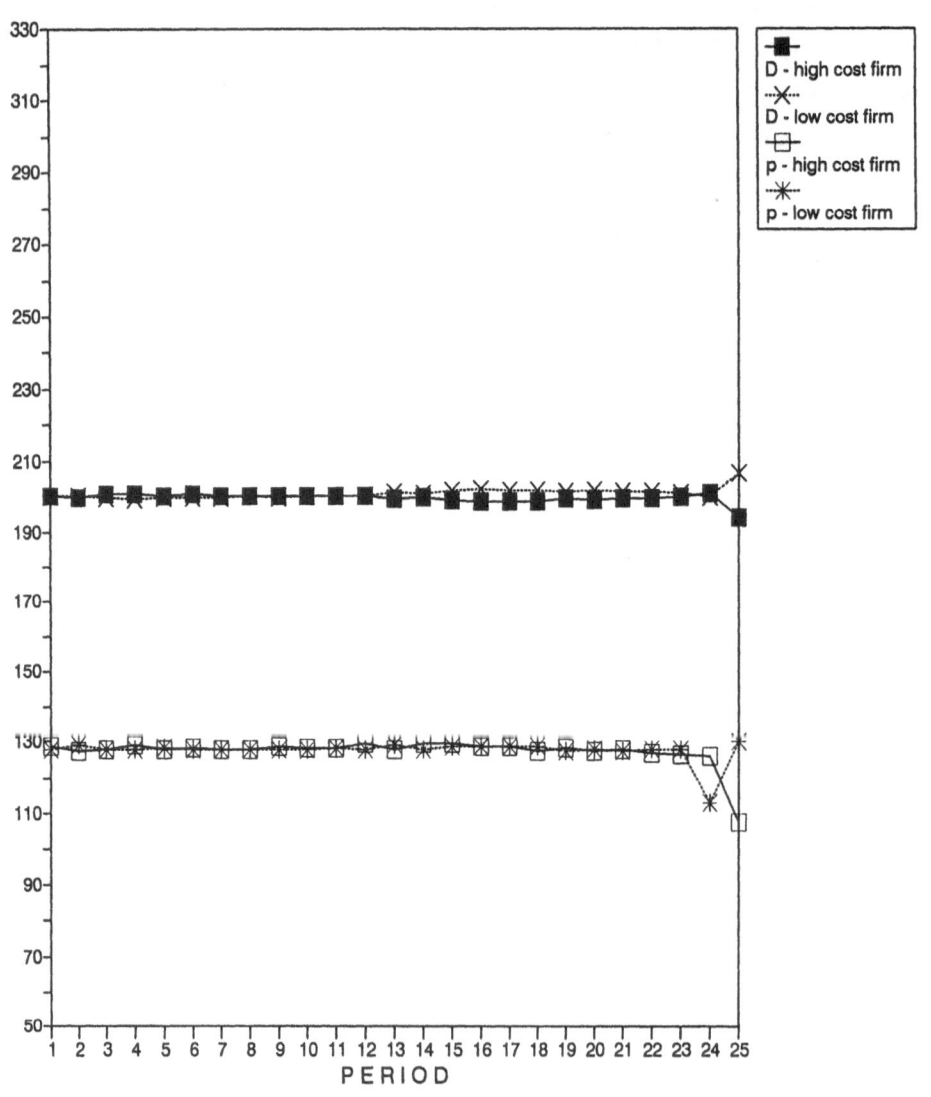

A WEAKLY COOPERATIVE MARKET

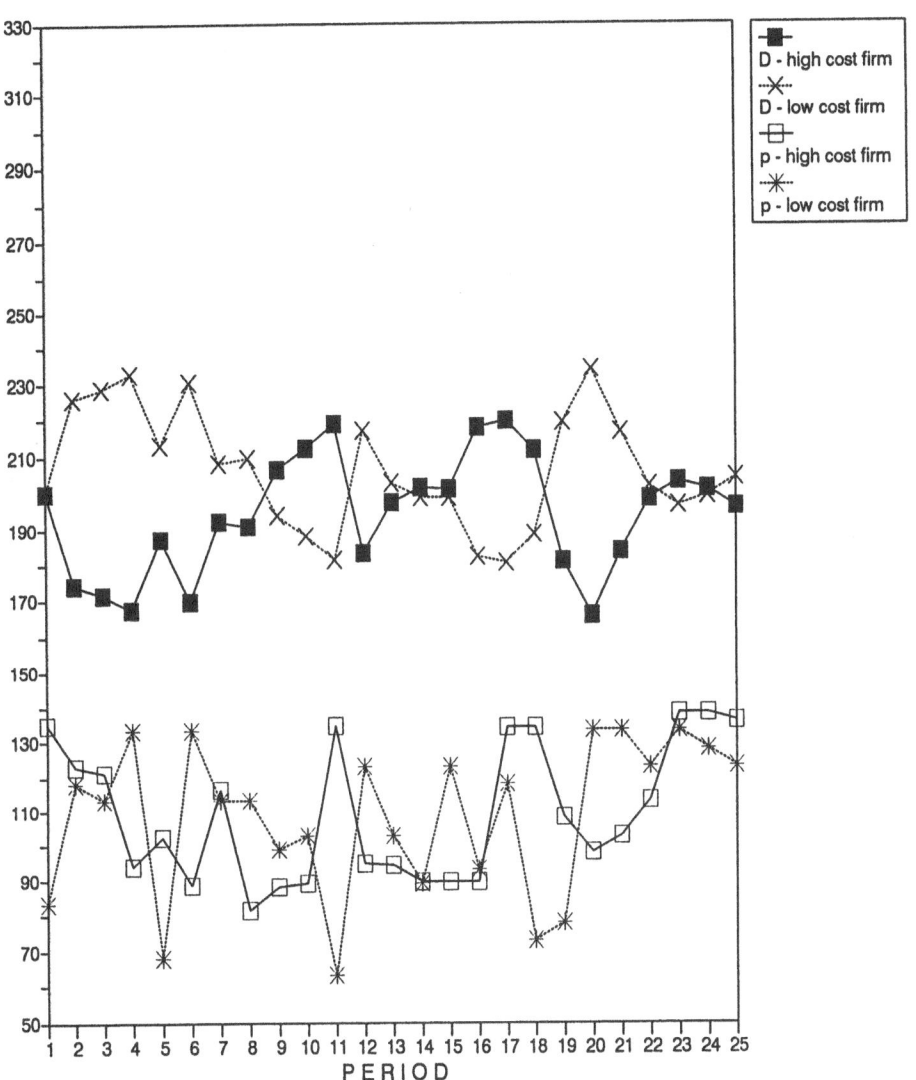

AN AGGRESSIVE MARKET

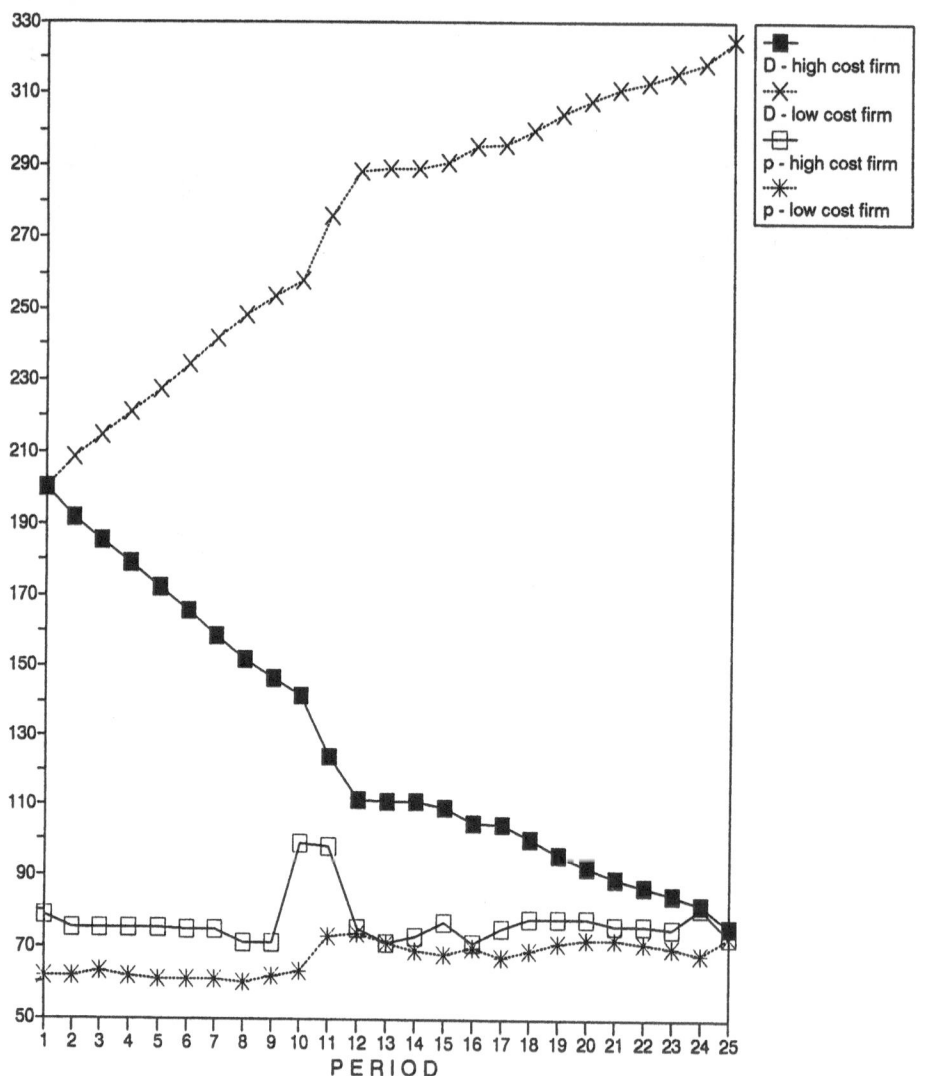

APPENDIX C

Flow charts of the strategies surviving in the evolutionary tournament

The participants of the strategy tournaments were asked to design the flow charts of their strategies such that they could be called up anew in each period.

The notation of the following variables used in the flow charts below was prescibed by the information brochure:

t	period
$D(t)$	own demand potential in period t
$D^*(t)$	opponent's demand potential in period t
$P(t)$	own price in period t
$P^*(t-1)$	opponent's price in period t–1

Some participants defined own variables within their flow charts.

LOW COST STRATEGY 18

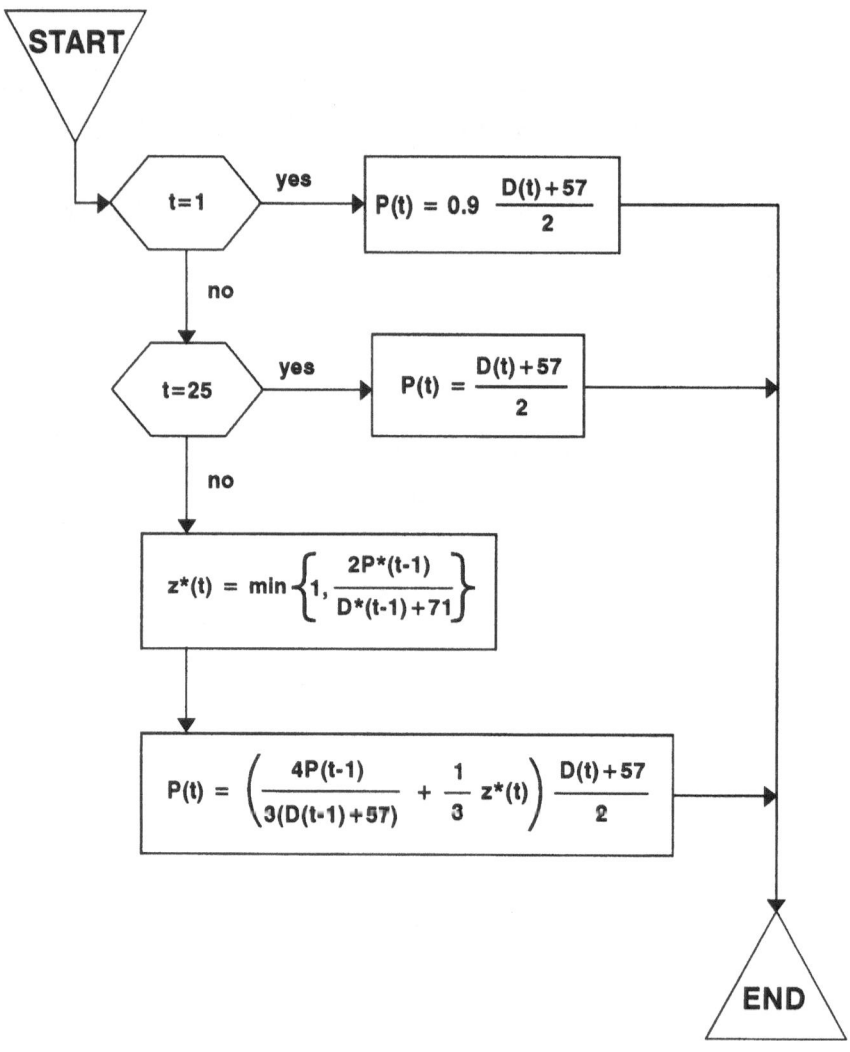

LOW COST STRATEGY 27

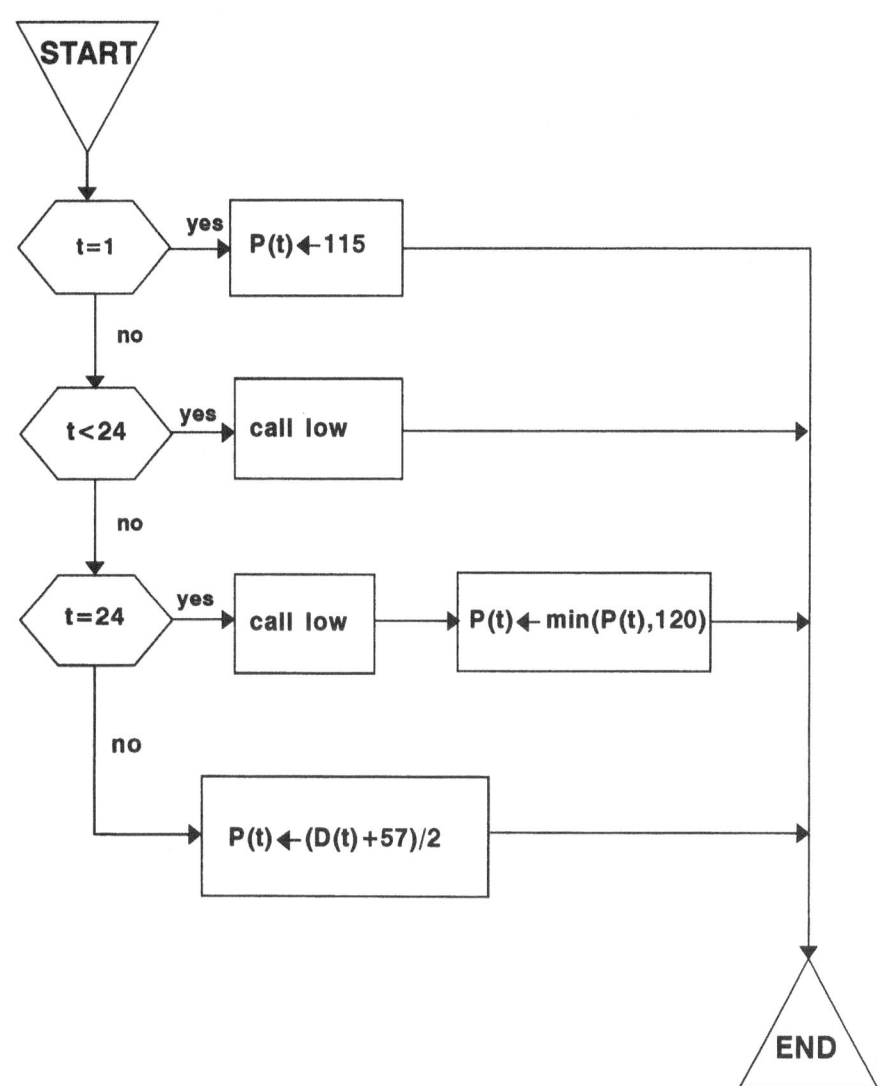

148

LOW COST STRATEGY 27: SUBROUTINE LOW

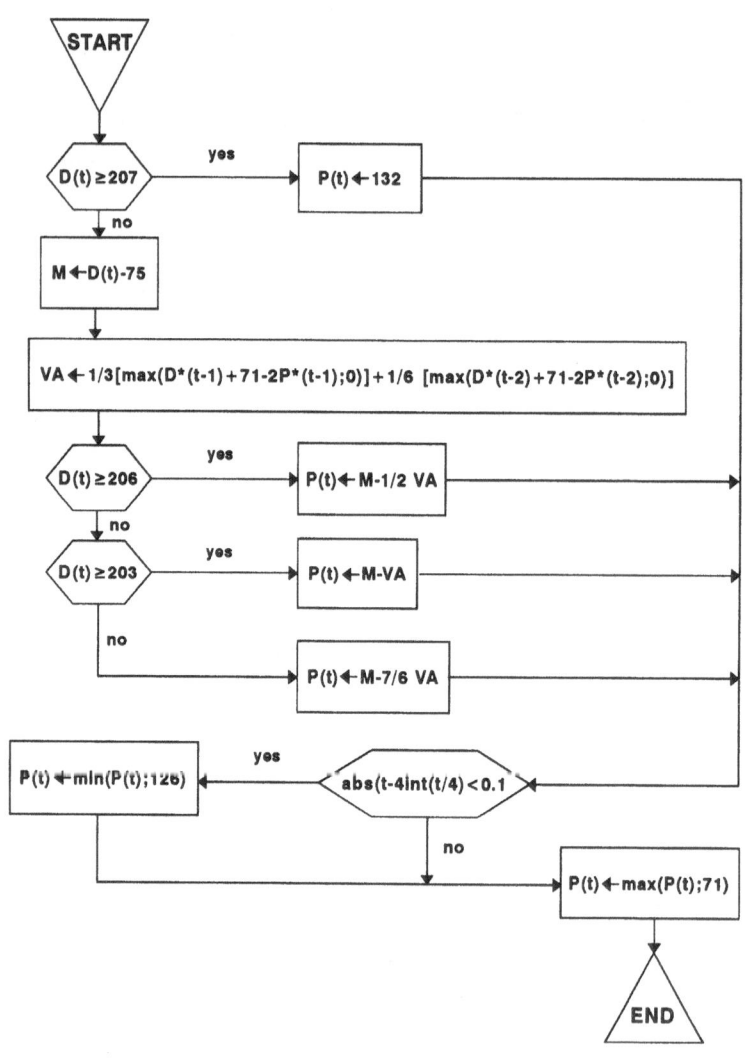

HIGH COST STRATEGY 17

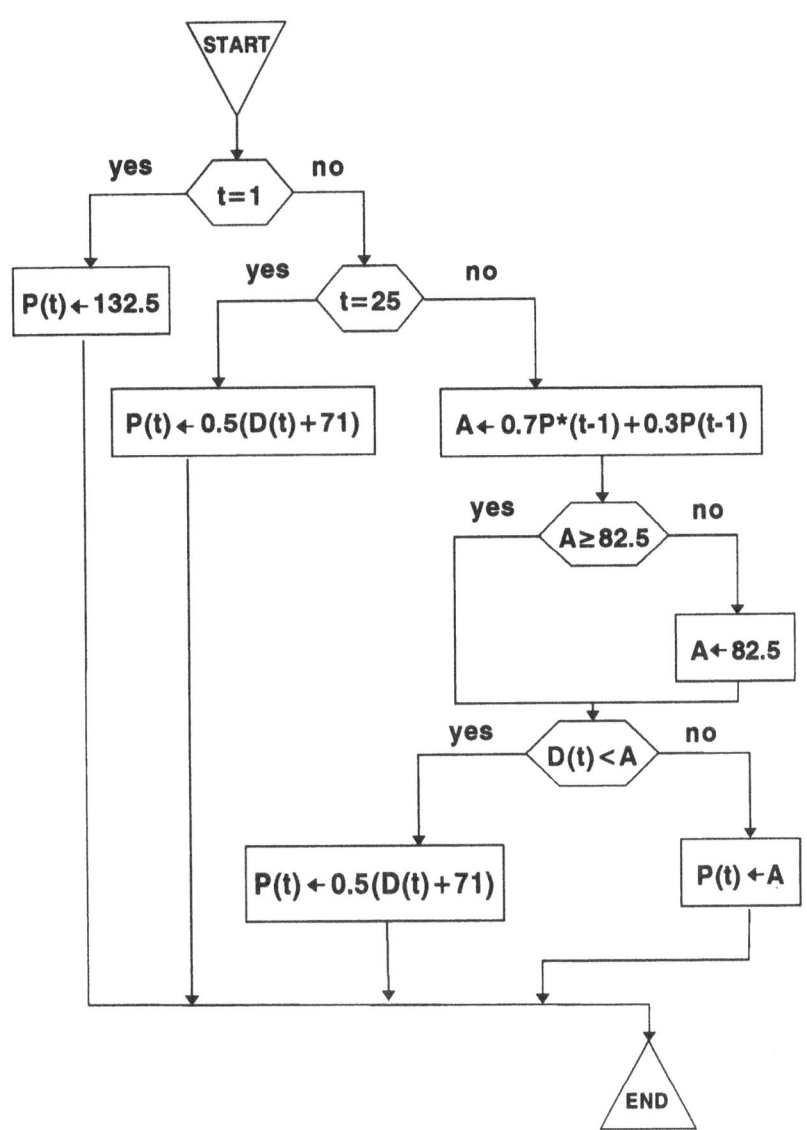

HIGH COST STRATEGY 31

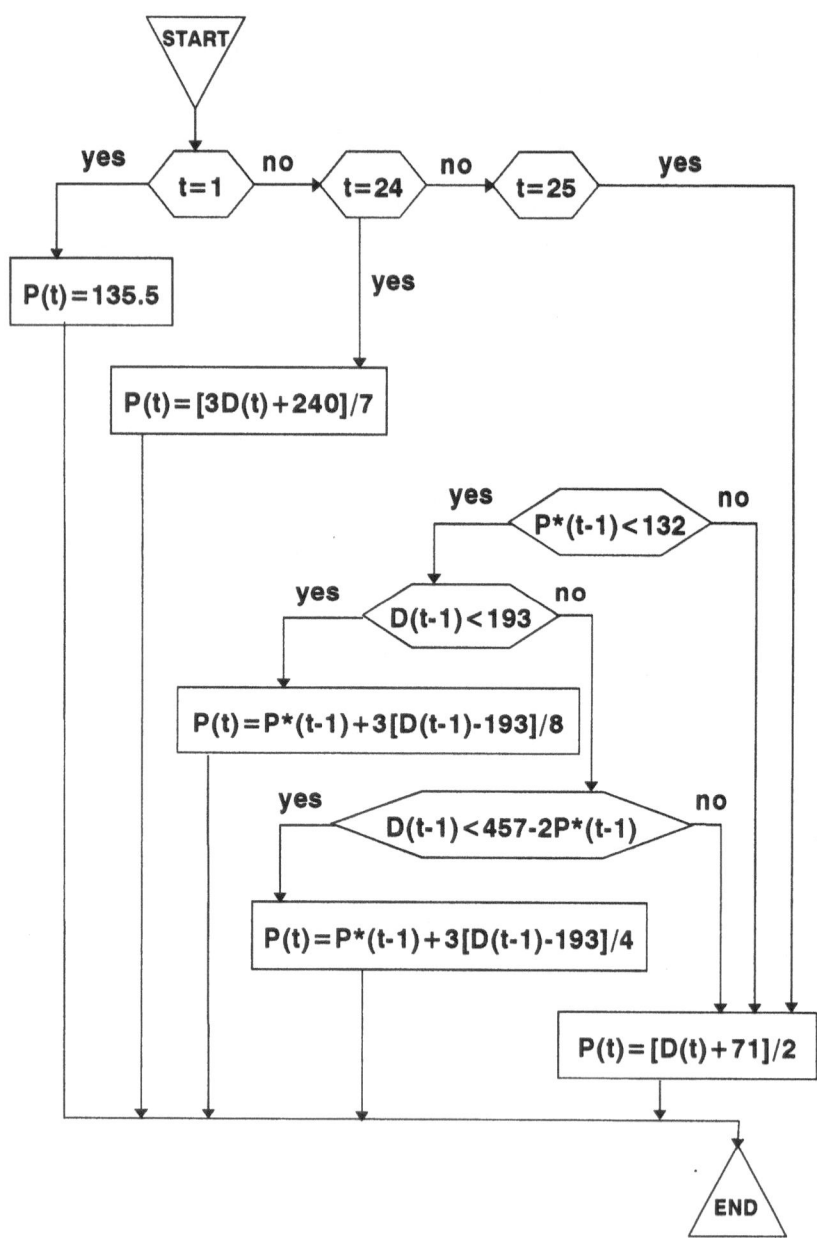

Lecture Notes in Economics and Mathematical Systems

For information about Vols. 1–210
please contact your bookseller or Springer-Verlag

Vol. 359: E. de Jong, Exchange Rate Determination and Optimal Economic Policy Under Various Exchange Rate Regimes. VII, 270 pages. 1991.

Vol. 360: P. Stalder, Regime Translations, Spillovers and Buffer Stocks. VI, 193 pages . 1991.

Vol. 361: C. F. Daganzo, Logistics Systems Analysis. X, 321 pages. 1991.

Vol. 362: F. Gehrels, Essays In Macroeconomics of an Open Economy. VII, 183 pages. 1991.

Vol. 363: C. Puppe, Distorted Probabilities and Choice under Risk. VIII, 100 pages . 1991

Vol. 364: B. Horvath, Are Policy Variables Exogenous? XII, 162 pages. 1991.

Vol. 365: G. A. Heuer, U. Leopold-Wildburger. Balanced Silverman Games on General Discrete Sets. V, 140 pages. 1991.

Vol. 366: J. Gruber (Ed.), Econometric Decision Models. Proceedings, 1989. VIII, 636 pages. 1991.

Vol. 367: M. Grauer, D. B. Pressmar (Eds.), Parallel Computing and Mathematical Optimization. Proceedings. V, 208 pages. 1991.

Vol. 368: M. Fedrizzi, J. Kacprzyk, M. Roubens (Eds.), Interactive Fuzzy Optimization. VII, 216 pages. 1991.

Vol. 369: R. Koblo, The Visible Hand. VIII, 131 pages.1991.

Vol. 370: M. J. Beckmann, M. N. Gopalan, R. Subramanian (Eds.), Stochastic Processes and their Applications. Proceedings, 1990. XLI, 292 pages. 1991.

Vol. 371: A. Schmutzler, Flexibility and Adjustment to Information in Sequential Decision Problems. VIII, 198 pages. 1991.

Vol. 372: J. Esteban, The Social Viability of Money. X, 202 pages. 1991.

Vol. 373: A. Billot, Economic Theory of Fuzzy Equilibria. XIII, 164 pages. 1992.

Vol. 374: G. Pflug, U. Dieter (Eds.), Simulation and Optimization. Proceedings, 1990. X, 162 pages. 1992.

Vol. 375: S.-J. Chen, Ch.-L. Hwang, Fuzzy Multiple Attribute Decision Making. XII, 536 pages. 1992.

Vol. 376: K.-H. Jöckel, G. Rothe, W. Sendler (Eds.), Bootstrapping and Related Techniques. Proceedings, 1990. VIII, 247 pages. 1992.

Vol. 377: A. Villar, Operator Theorems with Applications to Distributive Problems and Equilibrium Models. XVI, 160 pages. 1992.

Vol. 378: W. Krabs, J. Zowe (Eds.), Modern Methods of Optimization. Proceedings, 1990. VIII, 348 pages. 1992.

Vol. 379: K. Marti (Ed.), Stochastic Optimization. Proceedings, 1990. VII, 182 pages. 1992.

Vol. 380: J. Odelstad, Invariance and Structural Dependence. XII, 245 pages. 1992.

Vol. 381: C. Giannini, Topics in Structural VAR Econometrics. XI, 131 pages. 1992.

Vol. 382: W. Oettli, D. Pallaschke (Eds.), Advances in Optimization. Proceedings, 1991. X, 527 pages. 1992.

Vol. 383: J. Vartiainen, Capital Accumulation in a Corporatist Economy. VII, 177 pages. 1992.

Vol. 384: A. Martina, Lectures on the Economic Theory of Taxation. XII, 313 pages. 1992.

Vol. 385: J. Gardeazabal, M. Regúlez, The Monetary Model of Exchange Rates and Cointegration. X, 194 pages. 1992.

Vol. 386: M. Desrochers, J.-M. Rousseau (Eds.), Computer-Aided Transit Scheduling. Proceedings, 1990. XIII, 432 pages. 1992.

Vol. 387: W. Gaertner, M. Klemisch-Ahlert, Social Choice and Bargaining Perspectives on Distributive Justice. VIII, 131 pages. 1992.

Vol. 388: D. Bartmann, M. J. Beckmann, Inventory Control. XV, 252 pages. 1992.

Vol. 389: B. Dutta, D. Mookherjee, T. Parthasarathy, T. Raghavan, D. Ray, S. Tijs (Eds.), Game Theory and Economic Applications. Proceedings, 1990. ??, ?? pages. 1992.

Vol. 390: G. Sorger, Minimum Impatience Theorem for Recursive Economic Models. X, 162 pages. 1992.

Vol. 391: C. Keser, Experimental Duopoly Markets with Demand Inertia. X, 150 pages. 1992.

Vol. 392: K. Frauendorfer, Stochastic Two-Stage Programming. VIII, 228 pages. 1992.

Vol. 393: B. Lucke, Price Stabilization on World Agricultural Markets. XI, 274 pages. 1992.